FLYING WITH THE NAVY

DEDICATION

To my darling wife Heather, whose idea this book was and who guides me in all my literary endeavours

FLYING WITH THE NAVY

THE ROYAL NAVAL AIR SERVICE AND FLEET AIR ARM IN STUNNING RARE PHOTOGRAPHS

STEVE AND HEATHER BOND

GRUB STREET • LONDON

Published by
Grub Street
4 Rainham Close
London SW11 6SS

A CIP record for this title is available from the British Library

ISBN-13: 978-1-911714-03-3

Design by Myriam Bell Design, UK
Printed and bound by Finidr, Czechia

CONTENTS

PREFACE

DURING FIVE YEARS of research for a series of books on the post-war Fleet Air Arm, former FAA personnel not only provided us with a huge quantity of stories they also sent many thousands of photographs, largely taken by themselves. It soon became clear that what was needed was a photograph-heavy book showing off these stunning photos by giving them a full page, and quality reproduction. Whereas the other books were fascinating stories supported by photos, here you have fascinating photos supported by a small amount of explanation. It also seemed right to expand the coverage to cover the whole Royal Naval Air Service (RNAS) and Fleet Air Arm eras, in order to show better the changing role of the service throughout its history. Equally, it is just as important to show the 'people' side, not just the aircraft and ships.

Our grateful thanks go to all the people who so generously opened their photo collections to us, their names appear with their photographs – with a special mention for our good friend Dave Underwood who has accumulated a hugely impressive collection of pre-war photos. We must also thank Jane and Mike Cole-Hamilton, Bob Crane and Richard Dickinson, all Fleet Air Arm veterans, plus John Davies and his team at Grub Street for so enthusiastically offering to publish this book. We have strived to credit correctly all photographs and other material but the true provenance is not always identifiable with certainty. If we have omitted anyone, please accept our apologies.

Steve and Heather Bond
June 2023

FOREWORD

WHEN STEVE STARTED out on this project some five years ago, I am not quite sure he understood what he was taking on. The history of British naval aviation is of national importance and is an emotive subject which is supported by a hugely loyal and vocal group of supporters largely comprising current and retired Fleet Air Arm personnel and their families. As more and more came forward with their anecdotes, one volume quickly became four and we also now have this collection of fantastic photos covering over 100 years of naval aviation since the journey began in 1909.

If the flying machines provide the wherewithal to strike the enemy, then the life-blood of the Fleet Air Arm and indeed the Royal Naval Air Service beforehand, is the personnel who operate, maintain and support them. I have been privileged to serve most of my naval career of over 38 years with Fleet Air Arm personnel and it has been a rich journey to which I would sign up again tomorrow. Indeed, my association with this band of brothers (and sisters) continues to this day whereby I run our heritage charity of Navy Wings and am also chairman of the Fleet Air Arm Federation. Wherever I go, with either of these fine organisations, the people I meet always remind me of the esprit de corps, can-do attitude and above all the sense of humour which has provided the backbone and resilience for work in some of the most arduous and unforgiving conditions for aviation.

This is the final tome in this series of books and arguably it is the glue which brings them to life, with most of these photos having been taken by serving personnel at some stage in their voyage. I hope you enjoy this distinct collection of images and that it serves to remind many of you of your time in this very special, elite branch of the Royal Navy. For those of us who have lost friends during this period in uniform, and I suspect that will be most of us, then it also serves as a lasting memorial to their memory.

What a fine job the authors have done and, as we say in the Royal Navy, BZ Steve and Heather!

Commodore Jock Alexander OBE FRAeS
Chief Executive, Navy Wings

Left: *Yeovilton tower and a resident Merlin HC.4, 17 October 2019. (Robin Trewinnard-Boyle)*

INTRODUCTION

BRITAIN'S OTHER AIR FORCE FROM RNAS TO FAA

Somehow the Fleet Air Arm remains little known amongst the public at large. This is surprising for an armed service with such an illustrious history – its story is still being written today. The first launch of a fixed-wing aircraft, from a Royal Navy ship, took place on 10 January 1912. Lieutenant Charles Samson flew a Short pusher biplane off a platform constructed over the bow of the stationary battleship, HMS *Africa*. Following the success of further trials several merchantmen were converted to handle seaplanes operationally.

Army and naval aviation combined to form the Royal Flying Corps (RFC) in May 1912. The naval wing of the RFC operated until July 1914 when it became the Royal Naval Air Service (RNAS). In April 1918 the RNAS merged with the RFC to form the Royal Air Force. On 1 April 1924 the 'Fleet Air Arm (FAA) of the RAF' was formed. The FAA finally came fully under Admiralty control in May 1939.

During World War 2 the carrier fleet grew to reach a total of 59 by 1945, with an aircraft inventory of 3,700. After the war many of these aircraft, and the small escort carriers, were quickly disposed of. In the late 1940s four new carriers – HMS *Glory*, *Ocean*, *Theseus* and *Triumph* entered service. They were later joined by two large fleet carriers – *Eagle* in 1951 and *Ark Royal* in 1954; also four new light carriers, *Albion*, *Bulwark*, *Centaur* and *Hermes*. All the remaining World War 2 carriers except *Victorious* were withdrawn. The 1950s and 1960s saw more-capable aircraft enter service culminating in the Buccaneer and Phantom. Helicopters entered service, both on the carriers and on the rear decks of frigates and destroyers. Sadly in 1966 a planned new fleet carrier programme was cancelled by the government. Disposal of the existing fleet started that year, the last to go being *Ark Royal* in 1979.

However, with a continuing need for fleet air defence and anti-submarine work, three 'Harrier carriers' entered service in the 1980s – *Illustrious*, *Invincible* and a new *Ark Royal*, joined by a converted *Hermes*. These operated both Sea Harriers and helicopters, and later a joint RAF/FAA force flew the Harrier GR.7/GR.9 from them. The present-day Merlin helicopter arrived in 1999, and the Wildcat in 2013. Two new carriers – HMS *Queen Elizabeth* and *Prince of Wales* – were commissioned in 2017 and 2019. These embark F-35B Lightnings, Merlins and Wildcats. Since 1945 the FAA has seen action in crises and wars worldwide: Korea, Suez, Kuwait, the Indonesian Confrontation, Aden, eastern Africa, Rhodesia (Beira Patrol), Falklands, Gulf Wars, Balkans, Sierra Leone, Iraq/Syria, and Afghanistan. Today it remains an intrinsic part of our sea-borne capability. This is, of course, a great testament to all the men and women of the FAA, air and ground crew alike. They have taken everything thrown at them in their stride.

BEFORE AEROPLANES...

...there were balloons and airships

Householder (somewhat startled by descent of balloon which has not been quite the success it should have been). "So this is what you Navy Aircraft gentlemen do for a living!"

IN THE BEGINNING

Photos: Roz Himes collection

Right: *HMS Africa, a King Edward VII Class battleship. It served from 1905 until 1920 when it was scrapped.*

Below: *Charles Samson.*

Left: *On 10 January 1912 HMS* Africa *was used for the first test with shipboard aircraft, when Short-Sommer pusher biplane No. 2, flown by Lieutenant Charles Samson made the first take-off while the ship was at anchor in the River Medway. Later trials with the same aircraft and pilot were conducted off HMS* Hibernia. *The aircraft continued to be used for a variety of trials until it crashed on 28 January 1915 and was written off. A replica is on display at the Fleet Air Arm Museum.*

SHORT ADMIRALTY TYPE 74

Photos: David Rye (HMS Ganges Association)

THE SHORT ADMIRALTY Type 74 seaplane entered service with the RNAS in 1914; a total of 15 were acquired. Three took part in a raid on Cuxhaven on Christmas Day 1914; they flew from HMS *Empress*, *Engadine* and *Riviera*.

Below: *Type 74 serial No. 76 off RNAS Shotley seaplane station in June 1914, piloted by RFC naval wing CO Commander Charles Samson with Winston Churchill in the rear seat.*

Below: *The same aircraft at Shotley in 1914. No. 76 served at Grain and Calshot until, after overhaul by Blackburn's, it was wrecked at Dundee in February 1915.*

AVRO 504

The Avro 504's primary use was as a training aircraft; however, its moment of glory came courtesy of the navy. On 21 November 1914 four RNAS 504s bombed the Zeppelin airship works at Friedrichshafen, on the shores of Lake Constance. They were each carrying four 20-lb (9-kg) bombs. Although one aircraft was shot down the raid was successful. There were several direct hits on the airship sheds and the hydrogen-generating plant was destroyed.

The raid was mounted from Belfort in France and led by Squadron Commander E F Briggs who was shot down and taken prisoner. The other aircraft flown by Flight Lieutenant S V Sippe, Flight Commander J T Babington and Flight Sub-Lieutenant R P Cannon was forced to retire with a damaged tailskid.

Below left: *Avro 504A No.785, the variant used by the specially formed unit to attack the Zeppelin shed. (BAE Systems)*

Below right: *A later 504N K1974 of C Flight Gosport taxiing forward on HMS* Glorious *in 1935. (Dave Underwood collection)*

THE RNAS ON THE WESTERN FRONT

Photos: David Underwood collection

ALTHOUGH LITTLE REMEMBERED today, the RNAS contributed much to both the defensive and offensive aerial efforts on the Continent during the First World War.

Left: *Rare aeroplanes indeed. Just 15 of these Bréguet de Chasse pusher biplane fighters were acquired by the RNAS. Three of them are seen at 5 Wing Coudekerque in April/May 1916. Coudekerque is close to Dunkirk in the French department of Nord, Hauts-de-France, northern France. The type served from 1915 to 1917.*

Opposite page: *Seen in March/April 1916, two Caudron G.IV reconnaissance bombers, Nos. 3897 and 3899, of 7 Flight B Squadron 5 Wing are also at Coudekerque. A product of the French Caudron company, 55 examples were built for the RNAS and used, amongst other tasks, for attacks on German seaplane and airship bases in Belgium. They too only lasted in service for two years, from 1915 to 1917.*

SOPWITH CUCKOO

THE SOPWITH T.1 Cuckoo was the first landplane specifically designed for carrier operations. Deliveries to the Torpedo Aeroplane School at East Fortune commenced in early August 1918. Training took place in the Firth of Forth where Cuckoos launched practice torpedoes at targets towed by destroyers. Cuckoos of 185 Squadron embarked on HMS *Argus* in November 1918. No. 210 Squadron, who disbanded at Gosport on 1 April 1923, were the last to use the type.

Below left: *Cuckoo N6954, of 201 Training Depot Station East Fortune, dropping a torpedo sometime between July and October 1918. (Royal Canadian Navy [RCN] via John Bradley and Pat Martin)*

Below right: *Cuckoo N7192 on the Isle of Grain. (Steve Bond collection)*

SEAPLANES AND FLYING BOATS

Photos: Dave Underwood collection

THE FELIXSTOWE F.2A entered service in 1918, undertaking long-range anti-submarine and anti-Zeppelin patrols over the North Sea. On 10 May 1918 an F.2A, flown by Captains T C Pattinson and A H Munday, shot down Zeppelin L62 over Heligoland. The Norman Thompson N.T.4 boats, although lesser-known, flew patrols from the same air stations at Calshot, Cattewater, Dundee, Felixstowe, Invergordon, Killingholme and Scapa Flow.

Below left: *Short N.2B experimental seaplane N67 with the personnel of Test Flight Grain. The aircraft was stationed there from 1918 to 1920, however the Fairey III was chosen for service use instead.*

Below right: *Norman Thompson N.T.4A 'Little America' 9063 built at Middleton-on-Sea in June 1917. It was photographed at Grain's Type Test Flight where it arrived in July. Unfortunately it did not serve for long as it sprang a leak, and sank, whilst moored to a buoy at Westgate on 7 January 1918. It was written off after its wreck was towed ashore.*

Felixstowe F.2A N4462 on the River Medina at Cowes, Isle of Wight in early 1919. These aircraft were all built by Saunders (later Saunders and Roe) at East Cowes. This example was converted to an F.5, sadly its service record has not survived.

FAIREY FLYCATCHER

Photos: Dave Underwood collection

THE FLYCATCHER FLEET Fighter entered service in 1923. At sea they flew from the carriers HMS *Argus, Courageous, Eagle, Furious, Glorious* and *Hermes*; from 1924 until 1932 it was the only first-line fighter in service. It was also launched from gun-turret platforms on many capital ships. A total of 189 were built at Hayes, Middlesex, and the type was declared obsolete in 1935.

Below left: *A nice formation of Flycatchers, S1278 7, S1282 10, S1283 12, and S1286 14 of 408 Flight HMS* Glorious.

Below right: *Flycatcher N9662 5 of 402 Flight HMS* Eagle *over Malta in 1924. It survived in service until 1931.*

FAIREY SPOTTER-RECONNAISSANCE TYPES

Photos: Dave Underwood collection

Between 1927 and 1936 the Fairey IIIF served in every aircraft carrier. It was also used as a seaplane in 701, 714 and 718 Catapult Flights on capital ships and cruisers. In addition to many 400 series flights, the IIIF also equipped 820, 821, 822 and 823 Squadrons. After replacement in its primary role a small number lingered on in service until 1940.

The Fairey Seal was a development of the IIIF and was initially known as the IIIF Mk VI. It took over from the IIIF, entering service in 1933, with 821 Squadron in HMS *Courageous*. From 1936 it was progressively replaced by the Blackburn Shark; the last front-line units losing their Seals in 1937. It too stayed on in small numbers until 1940.

Below left: *Fairey Seal K3535 735 821 Squadron, HMS* Courageous*, taxiing with some major damage to its starboard mainplanes in late 1934.*

Below right: *Fairey Seals K3479 734, K3484 732 and K3535 735 from 821 Squadron HMS* Courageous *in 1934.*

Opposite page: *Already 'crewed up', Fairey IIIF S1404 801 of 823 Squadron, emerges from the hangar deck on HMS* Glorious.

S
140
901

BLACKBURN TORPEDO BOMBERS – DART and RIPON

OPERATING FROM HMS *Courageous*, *Eagle*, *Furious* and *Glorious*, the Blackburn Dart served from 1923. It was replaced by the Ripon from 1931 onwards. The last were withdrawn from 811 Squadron aboard HMS *Furious* in 1935. From 1929 it was largely succeeded by the Ripon which had greatly increased range; it thus required the addition of an observer to do the navigation! It too was withdrawn from service in 1935.

Below left: *Blackburn Ripon S1265 75 of 462 Flight, based on HMS* Glorious, *dropping a torpedo in 1929/1930. One hopes the merchant ship in the background keeps out of harm's way! (Steve Bond collection)*

Below right: *Blackburn T.2 Dart torpedo bomber N9629 69 of 460 Flight HMS* Eagle, *off Hal Far in 1925. It crashed into the sea on 6 August that year and was salvaged, but not repaired. (RNAS Hal Far Facebook)*

BLACKBURN TORPEDO BOMBERS – RIPON AND BAFFIN

THE RIPON SERVED with 460, 461, 462, 465 and 466 Flights plus 812 Squadron. It was a multi-purpose aircraft which was also used for long-range reconnaissance; when fitted with extra fuel tanks it had a maximum endurance of 14 hours – pretty tough for its crew!

Originally known as the Ripon V, the Baffin came along in January 1934 and was the first carrier-borne aircraft fitted with an air-cooled radial engine. Its performance was little better than the Ripon's and it served for just three years. Operating units were 810, 811, 812 and 820 Squadrons, flying from HMS *Courageous*, *Furious* and *Glorious*.

Below left: *Baffin K3548 71 of 812 Squadron being recovered after ditching off HMS* Glorious *on 3 March 1934. (Dave Underwood collection)*

Below right: *Ripon IIA S1563 15, S1559 10, S1560 14, S1567 16, S1562 12, and S1564 13 of 466 Flight HMS* Furious *in 1932/1933. (Squadron Leader F H Reed)*

HMS *COURAGEOUS* (D50)

HMS *COURAGEOUS* WAS commissioned in January 1917. Having originally been laid down as a battle cruiser she was converted into an aircraft carrier between June 1924 and May 1928. During a patrol in the Southwest Approaches, on 17 September 1939, signal was received from SS *Kafiristan* saying she was under fire from a German submarine. *Courageous* went to search for this ship or any survivors, and was hit and sunk by three torpedoes from U-29; 518 lives were lost.

Right: *A Fairey IIIF flying off. The type was flown off* Courageous *by 442, 445 and 446 Flights plus 820, 821 and 822B Squadrons. (Dave Underwood collection)*

Opposite page: *A sketch of Flycatchers flying over* Courageous *at Spithead in May 1932. (drawn by Midshipman David Cole-Hamilton)*

FLYCATCHERS, OVER COURAGEOUS
IN SPITHEAD

FAR-FLUNG PLACES – MALTA

Of all the overseas stations served by the RNAS and Fleet Air Arm Malta was probably the longest-lived. Air stations were situated at Hal Far, which opened as HMS Falcon in 1929, and Kalafrana used from 1917 to 1946. Hal Far survived as an air station until handed over to the RAF in 1965.

Below left: *The radio room at RNAS Hal Far in 1938/1939, with duty men Douglas Bensley Turner and J Clark looking pretty relaxed. (Maureen Teulier)*

Right: *Grand Harbour Valletta was used to seeing Royal Navy aircraft carriers. Local boatmen approach HMS* Ark Royal *in the 1970s. (Matt Deluchi)*

Fairey IIID N9739 of 481 Flight over the seaplane base at Kalafrana in 1925 or 1926. (Richard Forder)

HAWKER NIMROD

FROM 1931 THE much more advanced Hawker Nimrod replaced the Flycatcher as the standard carrier-borne fighter. Operationally it served with 401, 402, 404 and 409 Flights, plus 800, 802 and 803 Squadrons. The last to serve were with 802 in HMS *Glorious* in 1939.

Right: *Nimrod II K3660 573 of 802 Squadron Dekheila, Egypt in April 1939. On the 24th of the following month all Fleet Air Arm units were taken over from RAF control by the Admiralty. (Hugh Garlick archive via his son Mike)*

Opposite page: *Nimrod I Fleet Fighter S1580 572 802 Squadron from HMS* Glorious *looping over Malta in 1933. It survived with the unit until 22 January 1938 when it stalled in a turn and spun into the Mediterranean, killing its RAF pilot Pilot Officer Charles Henry Heber-Percy. (Steve Bond collection)*

HAWKER OSPREY

Photos: Dave Underwood collection

THE HAWKER OSPREY, the navy's first two-seat fighter, was developed from the Hart day bomber. The Osprey, fitted with the superb Rolls-Royce Kestrel engine, entered service in 1932; it was not until March 1939 that the last were withdrawn from front-line service. During that time it served with various flights and squadrons in HMS *Ark Royal*, *Courageous*, *Eagle*, *Glorious* and *Hermes*. It also served with 711, 712, 713, 714, 715, 716 and 718 Catapult Flights.

Below left: *Osprey S1688 205 of 407 Flight HMS* Exeter *2nd Cruiser Squadron 1932.*

Below right: *Osprey I K2782 286 of 405 Flight or 803 Squadron HMS* Eagle *at Gosport in 1932.*

CROWDED FLIGHT DECKS

Photos: Dave Underwood collection

Jack Routley flew Sharks in 1940: "I flew them at Worthy Down 755 NAS (Naval Air Squadron) to train telegraphist/air gunners – poor fellows. They practised their telegraphy and did some firing at drogues. We had an interesting variety of pilot mess-mates, including Laurence Olivier and Ralph Richardson."

Right: *Fairey IIIFs of 442 Flight on HMS Courageous off Cyprus in October 1931. Once again, every engine is running, ready for a mass launch as the carrier turns into wind.*

Overleaf: *Raring to go. HMS Courageous with 800 Squadron Nimrod fighters in the foreground, and 820 Squadron Blackburn Shark torpedo spotter aircraft to the rear, c.1935 (the Shark replaced the Baffin).*

105

FAIREY SEAFOX

Photos: Dave Underwood collection

THE SEAFOX WAS a fleet spotter carried by cruisers and entered service in late 1936. Strangely, whilst the pilot was in an open cockpit, his observer had the luxury of an enclosed one. In all 66 aircraft were built by Fairey's at Hamble with the last arriving in May 1939; its moment of glory came at the Battle of the River Plate in December 1939. One of HMS *Ajax*'s Seafoxes, flown by Lieutenant E Lewin and Lieutenant R Kearney, spotted throughout the action and on 17 December they reported that the pocket battleship *Graf Spee* had blown herself up, Lewin receiving the DSC for this action. The Seafox served on until July 1943.

Below: *K8587 H9A of 713 NAS on light cruiser HMS* Arethusa *in 1939. Delivered on 29 November 1937, this aircraft arrived at 713 Flight in HMS* Penelope *in the Mediterranean on 28 February 1938. The unit adopted squadron status in May 1939, and this Seafox continued to serve with it until that November.*

SUPERMARINE WALRUS – THE 'SHAGBAT'

THE WALRUS WAS a general-purpose amphibian which joined the Fleet in 1936, and was carried by battleships and cruisers for reconnaissance. It was also used extensively for air-sea rescue of downed airmen and even, occasionally, as a light bomber, including a raid on Italian Somaliland in November 1940. The type disappeared from use very quickly at the end of the war.

Right: *Walrus I W3085 B of 700 NAS launching from HMS* Gambia *in 1943. (Air Britain)*

Opposite page: *Walrus I K8544, seen here at Lee-on-Solent Pool, arriving after its first flight on 12 December 1936. It was shipped to Bermuda the following month. (Pat Martin collection)*

K8544

FAIREY SWORDFISH – THE 'STRINGBAG'

The action for which the Swordfish is best remembered came on the night of 11–12 November 1940. Twenty-one Swordfishes, led by Lieutenant Commander Kenneth 'Hooch' Williamson of 815 NAS, took off from HMS *Illustrious* to torpedo the Italian Fleet in Taranto harbour. Four warships were sunk and three more crippled. Only two Swordfish were lost, one flown by Williamson and Lieutenant Norman 'Blood' Scarlett who were taken prisoner, and the other by Lieutenant Gerald Bayly and Lieutenant Henry Slaughter who were sadly lost. Remarkably, for an aircraft introduced into service in 1936 and considered obsolete at the outbreak of the war, the Swordfish did not fly its final operational sortie until June 1945.

Below left: *The third Swordfish prototype K5662 on floats at the Marine Aircraft Experimental Establishment, Felixstowe in 1936 for manufacturer's trials. (Steve Bond collection)*

Below right: *Swordfish I K8386 945 heads a line from 824 Squadron HMS* Eagle *at Seletar in 1937–38. (Steve Bond collection)*

Fairey Swordfish II LS326, of what was then the Fleet Air Arm Historic Flight, landing at Duxford for the Flying Legends display on 13 July 2013; it now belongs to Navy Wings. The 'Stringbag' is probably the most famous FAA aircraft of all. (Steve Bond)

GRUMMAN MARTLET AND WILDCAT

THE MARTLET WAS the British version of the US Navy and Marine Corps F4F-3, and entered FAA serviced in September 1940 with 804 NAS. The later F4F-4B variant, with extra guns and some other changes, was supplied under Lend-Lease arrangements to the FAA where it was known as the Wildcat. These served widely especially in the Mediterranean and Far East theatres. Like most Lend-Lease aircraft all were quickly withdrawn after VJ Day.

Jack Routley flew Wildcats with 805 NAS in North Africa. "It didn't have the performance or manoeuvrability of the Seafire, but it did have .5 calibre machine-guns which were a better weapon than the .303 that the Seafires and Hurricanes had."

Right: *A colourised image of a Wildcat V, of 882 NAS HMS* Searcher, *in 1944/45. (James Smith archive via his son Kev)*

Opposite page: *This Wildcat belonged to 888 NAS and is seen on HMS* Formidable *in November 1942. Wildcats/Martlets served from 1940 to 1945. (Imperial War Museums via Getty Images)*

FAIREY BARRACUDA

THE BARRACUDA WAS a torpedo and dive bomber which, from January 1943, gradually replaced the Swordfish and Albacore. They first saw action in September of that year when 810 NAS supported the Allied landings in Salerno. On 3 April 1944 42 Barracudas dive-bombed the battleship *Tirpitz* scoring 15 direct hits. After serving in the Pacific theatre the Barracudas were mostly withdrawn from service by 1948. However 744 NAS hung on to theirs until May 1953.

Jim Langford remembers practice torpedo-dropping: "Climb up to 8,000 or 9,000 feet, stick her down in a 45-degree dive, 300–315 knots, and pull out at about 150 feet above the sea, aim towards your target ship, drop your torpedo and pull away."

Below left: *A rocket-assisted take-off, from HMS* Puncher, *for a Barracuda of 821 NAS in 1944–45. (Pat Martin collection)*

Below right: *A colourised image of a 769 NAS Barracuda landing on the wooden deck of escort carrier HMS* Searcher *(D40) in 1944 or 1945. (Kev Smith)*

Opposite page: *A Barracuda II of 825 NAS at RNAS Rattray, undergoing maintenance to its engine during August 1945. On top of the engine is AM(E) Martin Field, on the ladder is AM(A) George Lord and on the ground is AM(A) William Beattie. (Pat Martin collection)*

CHANCE VOUGHT CORSAIR

The Corsair proved to be one of the outstanding fighters of the Second World War. Entering US Navy service in 1943, the first Lend-Lease Corsairs for the FAA arrived that summer. It was perhaps best known for its operations in the Pacific where it acted both as fighter cover for Barracudas and Avengers, and as a strike aircraft. On 9 August 1945 Lieutenant Robert Gray of 1841 NAS earned a posthumous Victoria Cross for pressing home an attack on a Japanese destroyer, even though his aircraft was on fire. He lost his life when it plunged into the sea. The last FAA Corsair units, 1831 and 1851 NAS, disbanded in August 1946.

Below left: *Corsair being 'batted in' to land on HMS* Illustrious. *(Steve Bond collection)*

Below right: *The fate of so many aircraft in the Pacific Fleet after VJ Day. Unwanted Corsair III KD173, late of 1850 NAS, is pushed over the side into a watery grave from HMS* Speaker. *(Steve Bond collection)*

Opposite page: *Impressive propeller-tip vortices as a Corsair starts its take-off run on HMS* Searcher *in 1944–45. (Kev Smith)*

SUPERMARINE SEAFIRE

Photos: Patrick Martin collection

SEAFIRES SERVED IN many different marks from 1942 onwards. Their operational debut came in June 1942 when 801 and 807 NAS took part in Operation Torch, the invasion of North Africa. Subsequently Seafires were involved in all major FAA actions in the European and Pacific theatres. Despite the advent of jets, its operational swansong did not come until 1950, when 800 NAS flew Mk 47s on offensive patrols and ground-attack sorties from HMS *Triumph* during the Korean War.

Below: *Merlin-powered Seafire III NN574 B, of 803 (Canada) Squadron Nutts Corner, seen in July/August 1945. The Mk IIIs were replaced on the squadron by Mk XVs; all had left the squadron by December 1945.*

Opposite page: *Seafire XV PR479 L, of 803 (Canada) Squadron Nutts Corner, on HMCS* Warrior *in early 1946. The first of the mark arrived on the squadron in August 1945. The Mk XV was the first to be fitted with the Rolls-Royce Griffon engine.*

L
ROYAL NAVY
PR 473

CAPTAIN ERIC MELROSE 'WINKLE' BROWN CBE DSC AFC Hon FRAeS

'WINKLE' BROWN WAS both a Royal Navy officer and a test pilot. He flew no less than 487 different types of aircraft, more than anyone else in history. He also held the world record for the most aircraft carrier take-offs and landings (2,407 and 2,271 respectively). He made the first landing on an aircraft carrier by a twin-engined aircraft, the first by an aircraft with a tricycle undercarriage, the first by a jet aircraft and the first by a helicopter. He served from 1939 to 1970 and died in 2016.

Right: Captain Eric Melrose 'Winkle' Brown.

Opposite page: *The first jet aircraft ever to land on an aircraft carrier was this Vampire F.1 LZ551. Flown by 'Winkle' Brown it landed on HMS Ocean on 3 December 1945, as seen here. This same Vampire is now preserved in the Fleet Air Arm Museum at Yeovilton. (Steve Bond collection)*

AIRFIELD ON THE ROOF

The changing face of the aircraft carrier

Below left: *HMS* Argus *which served from 1918 to 1944, was the first example of what became the standard aircraft carrier layout. Converted from an under-construction ocean liner, she had a full-length flight deck and no obstructing island. (via David Rye)*

Below right: *HMS* Ark Royal *in 1957, showing the traditional 'cat and trap' carrier design albeit not quite in its final state, sailing with* Albion *and* Ocean*. The basic layout remains little altered today with the fully developed 'island', housing both the ship's bridge and 'FlyCo' (Flying Control), on the starboard side. (Steve Bond collection)*

CVS carrier HMS Ark Royal *tied up alongside, on the Elbe River Hamburg December 2010, showing her ski-jump for Harriers and Sea Harriers. This was her final port visit before decommissioning after 25 years of service. (Robin Trewinnard-Boyle)*

DE HAVILLAND TIGER MOTH

THE TIGER MOTH served with various air station flights, from 1939 until 1969. It was the quintessential pilot trainer, serving with the Fleet Air Arm in its traditional role, and as glider tug and general run-about. It was perhaps most associated with the Britannia Royal Naval College (BRNC) Flight at Roborough, giving naval cadets their first taste of flying. The Tiger Moth was finally replaced at BRNC in 1966 by the de Havilland Chipmunk.

Below left: *An unusual outing for a trio of BRNC Tigers aboard HMS* Eagle *on 11 July 1964 – arrestor wires and catapults not required! (via Mike Cole-Hamilton)*

Below right: *A cartoon by an unknown artist, drawn on* Hermes *to record a similar event in 1966. (Mike Cole-Hamilton collection)*

Opposite page: *Tiger Moth T.2 XL714 of Heron Flight Yeovilton at RAF Chivenor's open day on 23 August 1969. This was the last Tiger to remain in FAA service, finally being retired on 16 March 1972. (Adrian Balch)*

VL
XL714
ROYAL
NAVY

THE FIRST ROTARY-WINGS

Photos: via Jim Milne

Below: *Sikorsky Hoverfly I KL102 G of 771 NAS Worthy Down in 1946–47. An amusing set-up for the camera with the female pilot being handed a cup of tea. The Hoverfly was the navy's introduction to helicopter flying, and was simply used to gain experience with the concept.*

Opposite page: *Hoverfly I KK987 GJ-513, Hoverfly II KN879 GJ-501, and S.51 WB220 of 705 NAS Gosport sometime in January–March 1949. No. 705 NAS remained the helicopter pilot training unit until such training changed to tri-service. It is now one part of the Defence Helicopter Flying School at Shawbury.*

'An entertaining concept, but what possible military application?'

GJ 501
GJ-513
ROYAL NAVY

THE CHANGING FACE OF COCKPITS

AT FIRST ALMOST the only information available to the pilot in the cockpit was little more than airspeed, height, turn and slip and fuel content. Aircraft performance increased with technological improvement, and the cockpit became very complex and crowded. Today the situation has become one more related to information on demand, relieving the pilot of much of his cockpit-management workload.

Below from left to right: *A typical very basic Sopwith cockpit from the First World War. (Henry Parker); the more developed instrumentation in the Swordfish (Andy Vanes), and a plethora of dials and switches in the Sea Vixen. (Kim Sharman)*

The 'glass cockpit' of a commando Merlin HC.4. (Brian Marsh)

FAR-FLUNG PLACES – HOT AND COLD

Given the Royal Navy's remit, to be able to operate anywhere in the world, aircraft of the FAA have found themselves in many unusual places; often imposing demanding conditions for aircrew and maintainers alike.

Right: *A Walrus of 700 NAS at RNAS Boaz Island, Somerset, Bermuda. It seems to be having a spot of bother trying to come up the slipway. This seaplanes-only air station operated from 1939 to 1944. The slipway and the hangar, on the extreme right, are still there. (Rob Ricketts)*

Opposite page: *Lynx HAS.3 XZ233 ED-435, operating in the Antarctic from HMS* Endurance*, perched on the top of a passing iceberg on 21 March 1991 – just because they could! Pilot Lieutenant David 'Digs' Nelson and observer Lieutenant Brian Marsh, airborne between Anvers Island and the Lemaire Channel on the west side of the Antarctic peninsula. This was right at the end of the final deployment of the Falklands-era* Endurance*; the photo was taken from the ship's second Lynx. (Ken Morrison)*

FAIREY FIREFLY

THE FIREFLY WAS a highly versatile aircraft which served from 1943 to 1957. Operating initially in the fighter-reconnaissance role, it later added anti-submarine, night fighter, target tug and pilotless target drone to its armoury. It distinguished itself during the Second World War, particularly in the Far East where it flew from HMS *Illustrious*, *Indefatigable*, *Indomitable* and *Victorious*. It then added to its record during the Korean War as a strike aircraft, flying from *Glory*, *Ocean*, *Theseus* and *Triumph*, plus the Australian carrier HMAS *Sydney*.

Below left: *John Champion's first ever deck landing. Firefly AS.6 WD884, of 723 Squadron Royal Australian Navy based at Nowra, close to touch down on HMAS* Sydney *on 14 March 1956. Note the 'planeguard Sycamore in the distance. (John Champion)*

Below right: *The final Firefly, the AS.7, failed as an anti-submarine aircraft but saw a good deal of service as a pilot trainer. The last few lingered on as pilotless target drones. This one, WM767, was a drone shepherd aircraft handed over to the Royal Aircraft Establishment at Llanbedr and seen there in 1959. (Steve Bond collection)*

Opposite page: *Firefly NF.1 PP470 R-214 of the night-flying unit 'Black Flight' on Glory over Birżebbuġa, Malta. Flown by Lieutenant Tom Davies, with Ken Lambert down the back, it was en route to carrying out a flypast over Valletta for King George VI's birthday in June 1950. (Ken Lambert)*

R
214

SUPERMARINE SEA OTTER

THE SEA OTTER – a sort of 'grown-up' Walrus – entered service in late 1943. Flying in the same air-sea rescue and communication roles, it had a more conventional tractor engine layout and offered better performance and improved handling over its predecessor. The first operational unit was 1700 NAS at Trincomalee in Ceylon. Other units embarked in HMS *Ameer*, *Begum*, *Emperor*, *Khedive*, *Ocean*, *Theseus* and *Stalker*, with the last Sea Otters retiring from 744 NAS at Eglinton in January 1952.

Keith Cotton: "To start the Bristol Mercury engine two of us had to stand on the lower mainplane, insert a winding handle into the engine nacelle and, with some effort, wind up the inertia starter to full speed, shouting 'CONTACT' to the pilot who switched on the magnetos. We then pulled a toggle to engage the starter – the forward one of us was standing within just a couple of feet of the propeller when the engine started!"

Above: *Sea Otter ASR.II Ship's Flight aboard an unidentified carrier – possibly* Theseus *– in the late 1940s. (Air Britain)*

Opposite page: *A colourised photo of Sea Otter ASR.II RD894 of* Ocean *ship's flight in 1949. (Air Britain)*

O

GRUMMAN AVENGER

THE AVENGER TORPEDO bomber and anti-submarine aircraft joined the FAA in 1943. Originally it was called the Tarpon, but that name was quickly abandoned. As with other US Lend-Lease types the Avenger mainly earned its spurs in the Pacific Fleet, although it did have some notable successes against U-boats on the Arctic convoys. Although quickly withdrawn after the war the Avenger AS.4 to AS.6 re-appeared in the anti-submarine role from 1953 to 1955.

Jim Langford: "I joined 820 NAS on HMS *Indefatigable*. We went up to Japan, dropping bombs on aerodromes and shipping. The last raid by British Pacific Fleet was on the day the war finished, 15 August 1945. Sadly we lost one observer, and the Seafire squadron (887) lost one of their pilots on that last day."

Below left: *A colourised image of Avengers ranged at sunrise on the deck of an unidentified in carrier, 1944–45. (Kev Smith)*

Below right: *Another colourised image; an Avenger taking off from HMS* Searcher *in 1944–45. (Kev Smith)*

Opposite page: *Avenger AS.4 XB303 386 of 814 NAS Eglinton in 1954. The later ECM.6 was the last variant to remain in service, with only a handful lingering on with 831 NAS at Culdrose until October 1962. (John Coward)*

386

HAWKER SEA FURY

The Sea Fury is probably best known for its role in the Korean War. During that conflict Lieutenant Brian 'Schmoo' Ellis of 802 NAS shot down a MiG-15, the first victory by a piston-engine fighter over a jet. It was the last piston-engine fighter in front-line squadrons, entering service with 803 and 807 NAS in September 1947. In Korea 801 and 804 NAS were in HMS *Glory*, 802 NAS was in *Ocean*, 807 was in *Theseus*, and the Royal Australian Navy had 805 and 808 Squadrons in HMAS *Sydney*. From 1953 the Sea Hawk replaced the type in front-line service; the last to go left the fleet requirements unit at Hurn in 1962.

Right: *Sea Fury FB.11 VW717 of 802 NAS HMS* Vengeance, *with pilot Nick Cook and his ground crew ashore at Hal Far, Malta, early 1951. (Nick Cook)*

Opposite page: *Sea Fury FB.11 XV727 NW-114 of the Royal Australian Navy's 805 Squadron at Nowra, as flown by Dave Eagles in 1956–57 on his first loan service draft. (Dave Eagles)*

DE HAVILLAND SEA HORNET

Photos: Adrian Balch collection

THE SEA HORNET day and night fighter was a masterpiece, but sadly too late. Although it took piston-engine fighter performance to new heights, it was already being outclassed by the new jets. Consequently its service career was short, from 1946 to 1954, although a few lingered on with the fleet requirements unit at St Davids until 1957. Operationally it served with just two squadrons: 801 (HMS *Indomitable* and *Implacable*) and 809 (*Eagle* and *Indomitable*).

Peter Hiles: "At Ford we had a range of aircraft including the Sea Hornet which, of all the aircraft I flew, was my number one favourite aircraft. Funnily enough 'Winkle' Brown said exactly the same and, according to his logbooks, he flew 487 aircraft types and number one was the Sea Hornet. My unqualified comment is that it was a pure delight to fly – it was a splendid aeroplane. I flew it, but only for the joy of it on a few stooging missions."

Right: *The Sea Hornet F.20 prototype PX212 during deck trials on* Ocean *in April 1945.*

Opposite page: *Sea Hornet NF.21 VW957 J-481 of 809 NAS on* Eagle*, sometime between 2 September and 22 October 1953.*

481
J
380

WESTLAND WYVERN

A unique turbo-prop strike aircraft, the Wyvern served from 1954 to 1957 when it was replaced by jet types. The highpoint of its service was 830 NAS's (HMS *Eagle*) contribution to the Suez campaign in 1956. Wyverns also served with 813 (HMS *Albion* and *Eagle*), 827 (*Eagle*) and 831 (*Eagle*) NAS.

Jonathan Tod: "I never met a Wyvern pilot who hadn't ejected; I'm glad I was never asked to fly one! They used to do very steep dive-bombing during Suez in '56. It had two socking great contra-rotating propellers and they acted like a brake. As you released your bomb it then fell very neatly through the propellers! Whenever they were landing on it was 'Wyverns landing on, clear the flight deck!' The propellers were made out of forged steel; if they hit the flight deck, they just turned into shrapnel."

Right: *Wyvern S.4 WN326 J-375 830 NAS leading another to carry out a free take-off from HMS* Eagle *in 1956. (Derek Peskell via Martin Grant)*

Opposite page: *George Barras about to take a wire in his 830 NAS Wyvern S.4 WN328 J-371, HMS* Eagle *on 16 August 1956. (George Barras via his son David)*

SUPERMARINE ATTACKER

THE ATTACKER'S PLACE in history is secure as the FAA's first operational jet, however it can hardly be considered a great success. With a tailwheel undercarriage, not the best idea for a jet, and fitted with a very cumbersome long-range tank, its performance did not exactly sparkle. Nevertheless it was useful to help the FAA learn about jets. It served from 1951 to 1957, going to sea in HMS *Eagle*. The front-line units were 800, 803 and 890 NAS plus 1831, 1832 and 1833 RNVR Squadrons.

Nick Cook: "The aeroplane was ghastly, terrible. Can you imagine designing a jet with a tailwheel? When you were descending, with power off, it creaked like an old building. It was a real heap, and had some nasty characteristics. They had so many fatalities. Attackers on *Eagle* had fuel leaks into the plenum chambers and they just exploded with no warning. Guy Grantham, C-in-C Mediterranean, grounded them."

Right: *Attackers of 800 and 803 Squadrons on HMS* Eagle *in 1952. (Steve Bond collection)*

Opposite page: *Attacker F.1 WK320 838 of 1833 RNVR Squadron Midland Air Division Bramcote, seen at Honiley on 15 September 1956. (Steve Bond collection)*

SEA VENOM READY TO GO

Photos: Rob Faulkner

As a navalised variant of the RAF's Venom NF.2, the de Havilland Sea Venom was the navy's first true night/all-weather fighter. It entered service in 1954 and was the first to carry air-to-air guided missiles. The last Venoms left the front line in 1961 replaced by the Sea Vixen; although 831 NAS flew them in the electronic countermeasures (ECM) role until 1966. It also served on for some years in second-line roles until the air direction and training unit at Yeovilton retired the last examples in 1970. Sea Venoms flew extensive operations during the 1956 Suez crisis.

Above: *The aircraft handlers' proud badge – they were known as 'Roof Rats' due to their perilous work on the flight deck.*

Right: *Sea Venom FAW.22 XG612 V-461 of 893 NAS HMS* Victorious, *during early Blue Jay (later Firestreak) missile trials in December 1958.*

Seems there may be a problem as Sea Venom FAW.21 XG671 V-465 of 893 NAS sits ready to launch from HMS Victorious *in 1958, and a flight-deck crewman runs in.*

AIRCREW and GROUND CREW

ARTIFICER MIKE NORMAN had an interesting experience with a Sea Venom: "I manned the fire extinguisher as the Nene engine was started by cartridge with a bang and a cloud of black smoke. The aircraft was marshalled out, and we glanced at it as the engine was wound up for take-off. As it left the ground I saw one of the main wheels leave the aircraft! I was frozen to the spot mesmerised by the bouncing wheel, which was happily making its way towards the airfield perimeter fence!" The aircraft was flown by Jonathan Tod and was landed safely.

Right: *What would we do without the maintainers? The visiting aircraft team (VAT) of HMS* Ark Royal *Air Engineering Department stand ready to receive a visiting US Navy SH-60 Seahawk helicopter during the Auriga deployment of summer 2010. As air engineers, with a mix of fixed- and rotary-wing experience, they were called upon to apply chocks and lashings and host any visiting aircraft that didn't have their own ground crew. The only name known in this otherwise anonymous group is Junior Technician Jonathan Taylor, far right. (Robin Trewinnard-Boyle)*

Opposite page: *893 NAS pilot Rob Faulkner perched on the wing, and his observer Steve Lazenby on Sea Venom FAW.21 WW140 V-462 on HMS* Victorious *1958. Note the names of another crew, and the ground crew 'Plane Captain', by the air intake. (Rob Faulkner)*

EJECTION
SEATS
DANGER
DANGER
62
Lt. R. King
S Lt. P I Mathews
Lapm. B Watkins

A SHORE-BASED MISCELLANY

IT WOULD REQUIRE a book several times this size to cover every one of the different aircraft types that have been in service with the RNAS and FAA. Often overlooked are the many types shore-based for a wide variety of mainly secondary roles and which did not go to sea. Here is a small selection.

Below left: *Gloster Meteor T.7 WL332 HF-571 of 728 NAS Hal Far, at Decimomannu in March 1963. This Meteor variant was used both as a pilot trainer and a general 'hack' with station flights. (Jet Johnson via Pat Martin)*

Below right: *Another general duties type was the de Havilland Dominie, the military version of the Dragon Rapide. This one is X7488 serving with 782 NAS at Donibristle; the last few FAA Dominies lingered on into the early 1960s. (Steve Bond collection)*

The RAF and FAA set up the jointly manned 360 Squadron initially based at Watton in Norfolk, to provide an ECM capability; their operational aircraft was the English Electric Canberra T.17. WJ981 EN, seen here visiting Valley on 11 November 1983, by which time the unit was based at Wyton. (Steve Bond)

'BOOTNECKS' – THE ROYAL MARINES

THROUGHOUT THE ROYAL Navy personnel of the Royal Marines are referred to as 'bootnecks' or 'booties'. As far as the Fleet Air Arm is concerned Royal Marine commandos are part and parcel of the Commando Helicopter Force (CHF) at Yeovilton; the 'bootnecks' are frequently on board aircraft carriers ready to deploy to a trouble spot, wherever it may be. The first recorded use of FAA helicopters to insert troops into a combat zone was in the 1956 Suez Crisis.

The second and very different appearance of the Royal Marines, is when their band is carried on larger vessels including aircraft carriers, and the ship's deployment includes port visits, onboard receptions and ceremonials. Such events always end with the highly emotional playing of 'Sunset'.

Left: *The Royal Marines Band playing on retired HMS* Illustrious *during its final visit to Greenwich in May 2015. (Mike Garlick)*

Opposite page: *A CHF Merlin (with a Wildcat in the background) hovering behind its customers whilst parading for the public at Yeovilton's Air Day on 2 July 2016. (Adrian Balch)*

THE FLIGHT DECK CHALLENGE

Buccaneer pilot Jonathan Tod: "Most pilots approach their first night deck-landing with trepidation – a new skill to be mastered that tests flying abilities and psychological strengths. The view from the cockpit is unfamiliar – all the daylight visual marks have gone; you can't even see the ship until the last seconds. It is just a bunch of directional lights giving you all you need for a deck landing, centreline, glide path plus speed from your instruments. You are in the low wait as ATC spaces aircraft intervals down the glide path. As he is now in charge, you take a look at the instruments, your fuel state, aircraft weight, engines temperatures and pressures, radio frequencies, switch to 100 per cent oxygen, ejection seat in under water setting and tighten your straps.

"You are told to complete your landing checks, wheels, flaps and hook down, 'blow' on, airbrake out. Reduce speed to 150 knots and begin descent. Eyes down fixed on your instruments, whilst listening intently to the bleeps of your angle of attack/speed indicator. In about 90 seconds you will be hooked up on deck. In your headphones are the voices of your observer, the ATCO and the landing sight officer, who are all calm and determined to get 23 tons of aircraft onto a moving, pitching, ship – which you can't see! Each tells you of any minor deviation from the glidepath or centre-line and distance to run.

"Reduce speed to 135 knots – head up and see the bright centreline with dim lights each side of the landing strip. Concentrate, concentrate, precision, precision! Look left to the landing projection lights with the meatball in the middle of the green horizon bars indicating you are on the glidepath, keep it there. Speed must be accurate. You might catch a glimpse of the ship's navigation lights towering above you. Suddenly there is the thump of wheels on deck and a dramatic deceleration as your hook catches a wire. You have arrived! The more times you do it the more confident you become, but never relax as it can turn and bite you."

Left: *A time-lapse shot of an 800 NAS Buccaneer performing a night deck-landing on HMS* Eagle *in 1971. Note the sparks where the hook hit the deck. (John Eacott)*

Opposite page: *Landing can still be challenging by day! The pilot of Sea Fury FB.11 VR936 T-121 of 807 NAS is clearly struggling to put it down on HMS* Theseus*, 19 August 1948. (George Brewes)*

HMS FULMAR – RNAS LOSSIEMOUTH

LOSSIEMOUTH, ON THE Moray coast of Scotland, was taken over from the RAF in 1946 and commissioned as HMS Fulmar. It was returned to the RAF in 1972, although 849 NAS remained there until its Gannets were retired in 1979. Lossie was one of the principal fast-jet naval air stations and is perhaps best remembered as the long-term home of the Buccaneer fleet.

Below left: *736 and 801 NAS Buccaneers lined up for a flypast to commemorate the scrapping of HMS Victorious in 1967. (Steve Bond collection)*

Below right: *From the left Brian Toomey, Colin Casperd, 'Spiv' Leahy (CO), Tommy Leece (senior pilot) and Brian Wilson of the Sea Hawk FB.3 aerobatic team 'Red Hawks' 738 NAS summer 1957. (via Mike Cole-Hamilton)*

Opposite page: *Hunter PR.11 WT723 LM-692 of 764 NAS at Lossiemouth on 19 July 1969. (John Hughes via Adrian Balch)*

LM
692
WT 723

HMS GOLDCREST – RNAS BRAWDY

Brawdy in Pembrokeshire was taken over from the RAF on 1 January 1946 and commissioned as HMS Goldcrest II. It was later re-commissioned as HMS Goldcrest on 4 September 1952. It was returned to the RAF on 1 March 1971. The RAF left in 1992 since when Brawdy has been an army base called Cawdor Barracks.

Synonymous with fast-jet operations, Brawdy was home to many Sea Hawk squadrons before becoming a primary advanced training station with Hunters. It was also home to a naval aircraft support unit which looked after a variety of types including Sea Vixens.

Below left: *The Gannets of 849 NAS were at Brawdy from December 1964 until November 1970 when they moved to Lossiemouth. This is HMS* Eagle*'s COD.4 (carrier onboard delivery) XG790 rolling out at Brawdy on 1 June 1970. (Steve Bond)*

Below right: *Sea Vixen FAW.2 XJ581 of 899 NAS 'The Bunch with the Punch' taking off for an air test on 2 June 1970. (Steve Bond)*

Opposite page: *This Sea Vampire T.22 XA126 BY-557 bears the markings of home-based 727 NAS with whom it served before going into storage in 1960. Seen here at Brawdy's Open Day on 20 July 1963. (Steve Bond collection)*

557
ROYAL NAV
DHSEA VAMPIRE

HMS HERON – RNAS YEOVILTON

Opening in 1941 Yeovilton, in Somerset, for most of its time was looked upon as the FAA's fighter station. Its resident squadrons have flown such varied types as Seafire, Sea Venom, Sea Vixen, Phantom and finally the Sea Harrier. Its fighter days ended with the retirement of the Sea Harrier fleet in 2006. By that time the station already had a substantial rotary-wing community and, as the base for both the CHF and Wildcat forces, that is its major role today. It is also home to the Navy Wings historic aircraft fleet.

Below left: *Jonathon Whaley flying Navy Wings Sea Vixen FAW.2 XP924 en route from RAF Leuchars to Bournemouth. Sadly, it is now grounded following a later wheels-up landing. (Jonathon Whaley)*

Below right: *Sea Harrier FRS.1 ZD578 VL-715 of 899 NAS in the squadron's 50th anniversary marks. Taken near Yeovilton on 15 December 1992. (Adrian Balch)*

Opposite page: *Sea King HC.4 ZE427 K of 848 NAS over the Somerset levels on 11 February 2016 shortly before retirement. (Andy Aitchison)*

HMS OSPREY – RNAS PORTLAND

PORTLAND WAS A seaplane base between 1917 and 1919, and again briefly in 1939–40. In 1946 the first Hoverflies moved in and in 1959 it was commissioned as HMS Osprey. From then until it was closed in March 1996, Portland was a major helicopter station and home to Whirlwind, Wessex and Wasp units. Never really an airfield, however it did have a runway laid out with proper threshold markings, etc. – but it was just 755 feet long!

Below left: *829 NAS was the operational Wasp squadron providing helicopters for many small ships' flights. Here HMS Gurkha's Wasp is being batted in to land sometime in 1967–68. (Chris Brocklebank)*

Below right: *Wessex HAS.3 XS862 PO-650 of 737 NAS at Portland carrying a practice (non-nuclear) bomb c.1978–82. (David Paul)*

Opposite page: *Whirlwind HAS.7 XN308 PO-510 of 771 Squadron Portland, 1969–70. (Steve Bond collection)*

PO
DANGER
XN 308
ROYAL NAVY
ROYA' NAVY
510

HMS SEAHAWK – RNAS CULDROSE

THE NEWLY BUILT airfield at Culdrose, Cornwall, was commissioned and opened as HMS Seahawk in May 1947. Initially it housed the naval night-fighter school, soon followed by other fixed-wing units including 702 NAS, a jet-training unit, and 849 NAS with Skyraiders and later Gannets. Since 1958 it has been the FAA's principal helicopter station, today home to the Merlin ASW fleet. It still has a fixed-wing element with the Avengers of 750 NAS, the observer training unit. It is also the home of the School of Flight Deck Operations, formerly known as the School of Aircraft Handling.

Below left: *Mosquito TR.33 TW279 CW-413 of 790 NAS Culdrose in 1948. (Keith Lambert)*

Below right: *A tight fit! A Culdrose Sea King fills the flight deck on Type 22 frigate HMS Cumberland on 8 May 1990. (Nick Mayhew)*

Opposite page: *For many years 705 NAS was the FAA's helicopter pilot training unit. This Gazelle HT.2 XX436 is in the colours of the unit's 'Sharks' display team. (Adrian Balch)*

CU
39
ROYAL NAVY
DANGER

WESTLAND DRAGONFLY

Photos: Steve Bond collection

THE DRAGONFLY, DEVELOPED from the Sikorsky S-51, was the Fleet Air Arm's first truly effective helicopter and was a pioneer of helicopter air-sea rescue. From 1950 it could be found on most air station flights, and was a regular on ship's flights. It was largely supplanted by the Whirlwind in the early 1960s, although a few served on until 1967.

Tony Wilson: "It was fitted with a British Alvis Leonides engine. With its limited lifting capacity, it was the most impressive SAR aircraft, although it had already done valuable work in the Korean War and in rescuing civilians from the floods of the English east coast in 1953. It was, incidentally, quite a frightening machine to recover from a vortex ring state (a dangerous aerodynamic condition when a vortex ring system engulfs the main rotor), causing a severe loss of lift."

Right: *Dragonfly HR.3 WG750 J-901 of* Eagle *ship's flight at Gibraltar in 1953.*

Opposite page: *Dragonfly HR.5 WG709 977 of* Centaur *ship's flight in 1958–59.*

ROYAL NAVY
977

DOUGLAS SKYRAIDER

In November 1951 the first of 50 Skyraiders joined 778 NAS at Culdrose to introduce airborne early warning (AEW) for the fleet into service. No. 849 NAS became the sole operational unit and first took Skyraiders to sea in January 1953 in HMS *Eagle*. Thereafter separate flights served in all the fleet carriers, with those on *Albion* and *Eagle* in action during the Suez campaign. In 1960 the Skyraiders were withdrawn in favour of the Gannet AEW.3.

Right: *Skyraider AEW.1 WT959 J-415, and WV181 J-414, of 849 A Flight HMS* Eagle, *over Malta in June or July 1956. (Derek Peskell via Martin Grant)*

Opposite page: *The last ever Skyraider AEW.1 launch (actually a free take-off) by A-451 of 849 D Flight from* Albion, *in December 1960. (Mike Bickley)*

140
160
180

HAWKER/ARMSTRONG WHITWORTH SEA HAWK

THE SEA HAWK fighter and ground-attack aircraft entered service with 806 NAS at Brawdy in 1953. Six squadrons (800, 802, 804, 810, 897 and 899) took part in the 1956 Suez campaign with great distinction, flying from HMS *Albion*, *Bulwark* and *Eagle*. The type was finally retired in 1969. Designed and initially built by Hawker Aircraft, the company's 'Super Priority' Hunter production programme led to Sea Hawk production being handed over to Armstrong Whitworth after Hawker had built the first 35 aircraft.

Arun Prakash: "As much a pleasure to behold as a delight to fly, the Sea Hawk captivated me. The powered ailerons gave this straight-wing fighter good handling qualities, and the novice found her virtually viceless."

Left: *Sea Hawks on a 1950s officer recruiting poster.*

Above right: *806 NAS again, with a fine formation overhead* Eagle *in 1957. Note Wyverns and Sea Venoms on the deck. (Andy Copeland)*

Opposite page: *Sea Hawk FGA.6 XE408 A-167 of 806 NAS, taxiing to the catapult and spreading its wings on* Albion *in 1960. (Mike Bickley)*

STAND DOWN

Return to a more relaxed period of alertness

MIKE COLE-HAMILTON: "THE best was an alleged pipe during one of those endless patrols off East Africa in the mid-1960s. Over the tannoy 'There will be no Thursday this week. Thursday next week will be on Sunday as usual.'"

Below left: *A 'Sod's Opera' in* Eagle's *wardroom on January 1972. 'WAFUs' (Wet and Flippin' Useless – members of the FAA) to the fore. From left to right: Lieutenant Robin Cox, Flying Officer Lou Kemp RAF, Lieutenant Paul Collins, all 800 (Buccaneer) NAS. (Lou Kemp)*

Below right: *Mail from home.* Ark Royal's *quarter deck January 1972. (Gene Carolan)*

Opposite page: *A 'No-fly' Sunday on HMS* Ark Royal *in 1965. (Mike Garlick)*

FIRE

PERCIVAL SEA PRINCE

THE SEA PRINCE had an unglamorous, but much-valued, career as the principal tool for observer training, and as a light transport. It served from 1950 until May 1979 when 750 NAS finally phased it out in favour of the Jetstream.

Kim Sharman: "After Sea Vixens I flew Sea Princes with 750 Squadron at Lossiemouth. I was doing observer training – average to boring. But it was quite good for racking up the hours prior to leaving the navy, and you could also get your twin-engine rating. We used them occasionally for communications work to collect stuff or whatever. There was massive jockeying to do it as a change, 'You did it last time.' – 'I'm senior to you!' I thought, 'I can do this every day of the week if I want to', so I put my letter in and left the navy."

Below left: *An impressive formation of nine Sea Prince T.1s of 750 NAS (the Observer Training School), at Culdrose during their open day in July 1975. (Adrian Balch)*

Right: *Trainee observers in a 750 NAS Sea Prince T.1.*

Opposite page: *The 'Admiral's Barge', Sea Prince C.2 WJ348 belonging to 781 NAS as the private transport for Flag Officer Naval Air Command, seen in November 1953. (via Tom Singfield)*

SEARCH AND RESCUE

PROBABLY THE MOST public face of the FAA was their legendary search and rescue effort over so many decades. Whether it be a downed pilot to be winched out of the sea, a ship in distress or a holidaymaker stranded on the rocks, an FAA helicopter could be relied upon to save the situation. More recently all such SAR helicopter work has been contracted out to civilian companies.

Right: *The dramatic scene as an 814 NAS Sea King rescues the crew of the trawler* Ben Asdale *on 30 December 1978. Gale-force winds and driving snow meant the helicopter had to reverse towards the cliffs under guidance from the coastguard. After the rescue the* Ben Asdale *broke up. (Painting by John Hamilton via Mike Norman)*

Far right: *The Sea King crew were awarded, (left to right) pilot Tony Hogg – Air Force Cross, 2nd pilot Larry Jeram-Croft – Queen's Commendation for Valuable Service in the Air, observer Mike Norman – Air Force Cross, aircrewman Chris Folland – Queen's Commendation for Valuable Service in the Air. (Mike Norman)*

An 819 NAS Sea King carrying out a 4-hour 15-minute casevac (casualty evacuation) of an injured seaman from a Russian fish factory ship out past Rockall on 10 June 1974. Pilot Roy Swales, 2nd pilot David Paul, observer Trevor Dale, and crewman Colin Larcombe. (David Paul)

'CROSSING THE LINE' & A RUN ASHORE

'CROSSING THE LINE' is a centuries-old tradition whereby ship's company members crossing the equator for the first time are 'invited' into King Neptune's court and, regardless of their rank, they receive a ceremonial dowsing. The tradition dictates that the ship must pay its respects to the Lord of the Seas, King Neptune, to gain his acceptance. The ceremony involves sailors being tested to ensure they are capable of 'handling rough seas' and usually requires them to be painted and fed something with a revolting taste before being dunked!

Below left: *King Neptune's personal guards aboard* Centaur. *(David Millward)*

Below right: *A rather more relaxing 'run ashore' for 890 NAS Sea Vixen crews from* Ark Royal, *Mombasa 1966. From left to right: Stuart Sanderson, Mike Garlick, and Roy Kenward. (John Bradley)*

Opposite page: *'Crossing the Line' in full swing on HMS* Ark Royal *in 1965. (Mike Garlick)*

DE HAVILLAND SEA VIXEN

Sea Vixens served as both fighters and strike aircraft from 1959 to 1971. They flew from HMS *Ark Royal*, *Centaur*, *Eagle*, *Hermes* and *Victorious*. Units included 766, 890, 892, 893 and 899 NAS and finally the Fleet Requirements Air Direction Unit at Yeovilton. Sadly the Vixen had a poor safety record, and during its service no fewer than 51 aircrew and at least four ground crew lost their lives in accidents.

Right: *Observer Flying Officer Dave McNamara RAF and his pilot Lieutenant Paul Chaplin in front of the SNEB rocket pod fitted to their 899 NAS Sea Vixen FAW.2, pre-departure on* Eagle *in 1970. (Paul Chaplin)*

Opposite page: *Sea Vixen FAW.1 XN707 of 890 NAS* Ark Royal *in 1963 or 1964. (Mark Thomson)*

R
ROYAL NAVY
XN707
242
XN707

SUPERMARINE SCIMITAR

THE SCIMITAR WAS a fighter and a nuclear-capable strike aircraft. It was probably the most challenging type to fly from HMS *Centaur* which was, frankly, far too small for high-performance jets. The type served with 700, 736, 764, 800, 800B, 803, 804, 807 NAS aboard *Centaur*, *Eagle*, *Hermes* and *Victorious*.

Brent Owen flew Scimitars from *Ark Royal*: "Frequently we would be in the aircraft on the catapult and tensioned, only to have the launch cancelled. When tensioned on the catapult the tail of the Scimitar was on the deck, with the cockpit high and reclined. Quite comfortable in a sense, enough on occasions to nod off until someone banged on the side of the fuselage. I developed the self-preservation habit of refusing to retard the throttle, after a cancelled launch, until the catapult officer stood in front of the aircraft."

Right: *Scimitar F.1 XD268 of 800B NAS just leaving the deck on* Eagle *1964. 800B had the role of air-to-air refuelling the squadron's Buccaneers S.1s. The Buccaneers frequently had to launch without full fuel loads, due to their lower powered Gyron Junior engines. (Mark Thomson)*

Opposite page: *Scimitar F.1 XD235 of the Fleet Requirements Unit near Hurn on 5 June 1968. (Adrian Balch)*

ROYAL NAVY
XD235
032

PHANTOM SUNSET

The McDonnell Phantom FG.1 was, without doubt, the highest performance aircraft to fly from the traditional 'cat and trap' carriers. It was in service from 1968 to 1978. Although carrier trials were conducted on HMS *Eagle* in 1969, operationally it was only *Ark Royal* which took them to sea. No. 766 NAS was the Phantom training unit (later replaced by the Phantom Training Flight), and 892 NAS was the operational unit.

Gene Carolan: "I took the photo, the pilot of the aircraft you see was Tim Gedge."

Tim Gedge: "Gene and I compared logbooks and decided that the photo was taken on 24 February 1970. He was the pilot of Phantom XV586 with Sub Lieutenant Des Hughes as the observer. I was the pilot of the aircraft in the photo, XV590 with Lieutenant Pip Coombes as my observer. Both our logbooks show the sortie as VHLAI (very high level air intercepts) lasting just one hour. My entry shows Sennan/350 LNZ 95, which indicates we were operating from Land's End on the 350-degree radial out to 95 nm."

Below left: *Phantom FG.1 R-011 892 NAS landing on* Ark Royal. *(Phil Boyden via Richard Andrews)*

Below right: *Phantom FG.1 XV587 R-005, 892 NAS en route from* Ark Royal *to Yeovilton 15 December 1970. (Gene Carolan)*

Opposite page: *An 892 NAS Phantom photographed from another. (Gene Carolan)*

PEOPLE – OUT OF THE ORDINARY

JOHN RAWLING: "THE general service elements of the navy, seamen etc. never regarded us FAA as 'real sailors'. What always amused us was the continued use, in a modern navy, of many of the traditional technical terms for procedures and bits of kit. Phrases such as 'heave away handsomely', 'marry the falls', and 'close up those halyards' always amused us. Many a painter was still 'lashed to a lizard' and many a line was 'flemished down'.

"To annoy these sea dogs we used to make up terms such as, 'flange the scransoms', 'stinge the futtocks' etc., our imagination knew no bounds, and much fun was had. When in their presence we also annoyed our salt-encrusted brethren by referring loudly to floors instead of decks, ceilings instead of deckheads, walls instead of bulkheads, and drains instead of scuppers. Our favourite though was to refer to portholes, or worse, scuttles, as windows. We all had much fun winding up the 'fisheads'."

Below left: *HMS Albion's 2n44mess [multi-man cabin] with, from left to right, Aircraft Handlers Mick Crowson, Tommy McKie and 'Blacky' (how could it be anything else?) Blackman. (Chris Pugsley)*

Below right: *The control room in HMS* Invincible *during an exercise. Unfortunately, the rating's name is unknown. (Steve Bond collection)*

Opposite page: *'Hands to washing aircraft'. An 801 NAS Buccaneer gets the treatment on HMS* Hermes *in the Aegean Sea May 1970. (David Gearing)*

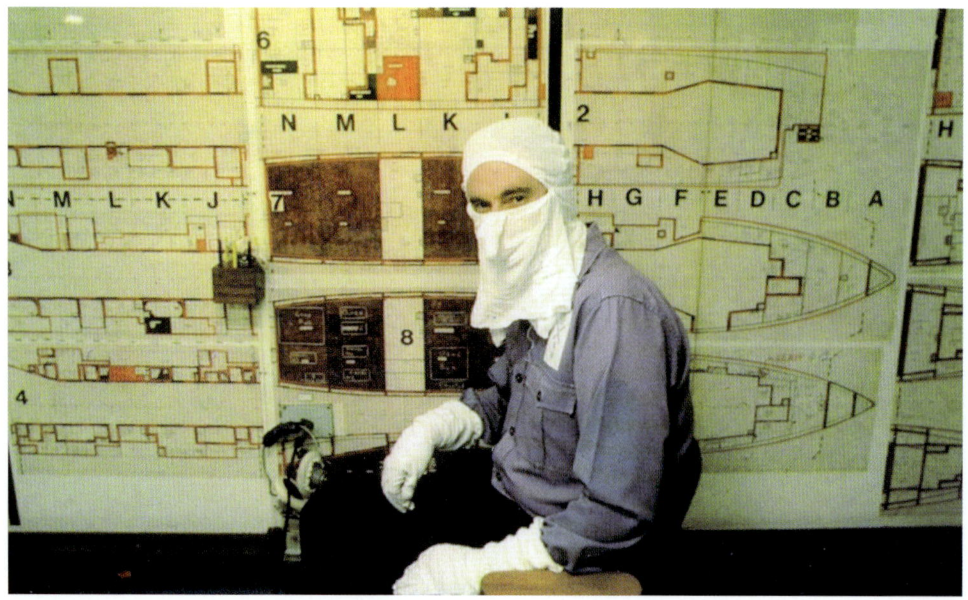

YAL NAVY
XV162

WESTLAND WASP

BASED ON THE Saunders-Roe P.531, the Wasp was the FAA's first helicopter specifically designed to be operated from destroyers and frigates. It entered service in the anti-submarine role with 700W NAS at Culdrose in 1964. They were also flown by 703, 705, 706 and 771 NAS whilst the operational unit, which detached flights to individual ships, was 829 NAS based at various times at Culdrose and Portland. The last Wasps were withdrawn from service in 1988; its replacement was the Lynx.

Larry Jeram-Croft: "Thank God I never went to sea in a Wasp; in an underpowered machine like that, at night. I know three pilots who went up to the bridge, after a night landing, and told the captain they were never going to do a night flight again. It was pretty marginal."

Right: *Wasp HAS.1 XS544 456 829 NAS, HMS* Yarmouth *ship's flight, Keith Copus flying the ensign during a display in 1976. (Keith Copus)*

Opposite page: *829 NAS Wasp HAS.1 XS545 452, of HMS* Galatea*'s ship's flight, ashore on Aldabra Atoll in the Seychelles, 1970. (Jeff Peck collection)*

XS545 ROYAL NAVY
452
DANGER

THE MIGHTY BUCCANEER

BLACKBURN'S MASTERPIECE – the ultimate low level strike aircraft. The Buccaneer served with 700, 736, 800, 801, 803 and 809 NAS and went to sea aboard HMS *Ark Royal*, *Eagle*, *Hermes* and *Victorious*. First deliveries in 1961 were of the notoriously underpowered S.1 with the Gyron Junior engine. Everything changed when the superb Spey-powered S.2 came along in 1965 and served through until the withdrawal of *Ark Royal* at the end of 1978.

Chris Pugsley: "I remember a Buccaneer doing a fly-by down the port side at dusk. It was as solid as a rock at high speed, low level and the sight of it doing twinkle rolls as it passed the ship was awesome."

Right: *Doing what a 'Bucc' did best! An 809 NAS S.2 XT287 R-033 from* Ark Royal *low over the sea; as seen from RFA* Lyness *in 1978. (Graham Collins)*

Opposite page: *Buccaneer S.2 XV162 H-232 of 801 NAS ready to launch from* Hermes*; April 1970 in the eastern Mediterranean. (David Gearing)*

232

GANNET ON THE 'CAT'

THE GANNET WAS initially produced as an anti-submarine aircraft. It entered service in April 1954 with 703 NAS at Ford. Thereafter it went to sea with multiple squadrons aboard HMS *Albion*, *Ark Royal*, *Bulwark*, *Centaur* and *Eagle* until retiring from operational use in 1960. In 1959 the AEW.3 arrived to replace the Skyraider. It served with 849 NAS (in *Ark Royal*, *Centaur*, *Eagle*, *Hermes* and *Victorious*) until the final withdrawal of *Ark Royal* at the end of 1978.

Martin Davidson: "You couldn't beat the first flight of the morning on the *Ark*, the smell of aviation fuel, the bakery vent with smell of fresh bread baking and the anticipation of breakfast. The chill of the breeze on your face as the ship turns into the wind, the steam from the cat, the heat off the afterburners as the first 'heavies' take off. The first Gannet getting ready, the heat of the Palouste pipe, the draft off the props, the 'cab' straining against the strop on the cat, the FDO counting it down, the noise of the double mamba going full chat. I loved those days."

Right: *Gannet AEW.3 XL479 R-044 of 849 NAS B Flight on* Ark Royal*'s bow catapult in the late 1970s. The Palouste starter sits ready, and the aircraft is still chained to the deck awaiting its crew. (Tim Woodman)*

Opposite page: *Anti-submarine Gannets. A line of 719 NAS AS.1s go through the amusing wing-spread routine whilst taxiing out at Eglinton in 1957. No. 719 was the Naval Anti-Submarine Operational Flying School. (John Coward)*

DE HAVILLAND SEA DEVON AND SEA HERON

Photos: Stephen Wolf

When the FAA was much larger than it is today it had a good number of air stations, so it made sense to have a dedicated fleet of communications aircraft. For many years that task fell on two de Havilland types, the twin-engine Sea Devon and the four-engine Sea Heron. Both served with station flights, as 'Admiral's Barges', and a number of squadrons, primarily 781 NAS at Lee-on-Solent. The Sea Devon served from 1955 to 1988 (its last unit being 771 NAS at Culdrose), and the Sea Heron from 1961 to 1989 (lastly with Heron Flight Yeovilton).

Right: *Sea Devon C.20 XJ319 an 'Admiral's Barge' from Station Flight Yeovilton, landing there on 7 September 1973.*

Opposite page: *781 NAS Sea Heron XR443 in formation with a Sea Devon of the same unit over Lee-on-Solent on 22 July 1978.*

ROYAL NAVY
XR443

'PAF' AND 'TUGG'

EXAMPLES OF THE work of the Fleet Air Arm's two renowned cartoonists. Below, Lieutenant Cecil 'Tugg' Willson MBE pays tribute to all the many trades involved in the FAA. Opposite, Lieutenant Commander Peter 'PAF' Grant MBE and some of his many cartoons focussed on 849 NAS with which he served. 'Tugg' was extraordinarily prolific with his cartoons, and both men were endowed with a wicked sense of humour.

QUEEN ELIZABETH II'S SILVER JUBILEE REVIEW JUNE 1977

TO CELEBRATE THE Silver Jubilee of Her Majesty Queen Elizabeth II, a Royal Review of the Fleet was held at Spithead. In addition to many Royal Navy warships, there were 14 foreign navies represented, plus 81 merchant navy and other vessels. The planned FAA flypast was badly affected by the weather which caused the cancellation of the fixed-wing element.

Scott Wilson: "The weather on the day of the Queen's Review turned out much worse than forecast. We headed out of Lee for an anti-clockwise circumnavigation of the Isle of Wight, 100 feet above the sea. I logged an hour and 40 minutes of tense formation flying, and was in and out of cloud a few times, before we finally paid our respects to Her Majesty."

Below left: *Pilot's eye view taken by Scott Wilson as he passed over the Royal Yacht* Britannia, *with* Ark Royal *in view.*

Below right: *Taken on rehearsal day when the weather was fine. HMS* Ark Royal *is being overflown by a large formation of Wessex Mks 1 and 3 operating from Lee-on-Solent, with a Lynx and Gazelle formation in the distance. (Phil Chadwick)*

FAREWELL *ARK ROYAL*

LAID DOWN IN May 1943 in Cammell Laird's shipyard, HMS *Ark Royal* was commissioned in February 1955. The cancellation of a new-build carrier programme in the mid-1960s meant that, by the time of her retirement, she was the last fixed-wing carrier still serving with the Royal Navy. In September 1980 she was towed to Cairnryan where scrapping commenced the following month.

Below left: Ark Royal*'s earlier captain, Wilfred Graham, and 'Little Wilf' the notorious star of the TV series 'Sailor' in 1976. 'Little Wilf' retired to Codnor Inn in Derbyshire but now lives in sunny Spain.*

Below right: *On 4 December 1978, in the cold light of a winter's morning, the Ark, arguably Britain's most famous aircraft carrier enters Devonport for the last time under the command of Captain 'Ted' Anson. Despite the early hour a large crowd has gathered to wave her in, as she flies her paying off pennant. (Dave Watts)*

STAND DOWN

Top right: *Catherine Davies: "Our Yeovilton ship's daily orders had featured a cartoon character called 'Phantom Phred'. I had joined the canoe club and one summer he acquired a cartoon girlfriend 'Phantom Phemale', with news of her canoe club events. A decision was made to build a model Phantom about 14 feet long. We went to water carnivals with the Phantom mounted on a canoe and towed by the lads, with me perched on top wearing a 'Fly Navy' banner. On a lovely summer evening we floated along the River Avon, in Stratford-on-Avon, past the theatre where the cast of the current Shakespeare play were on the balcony during their intermission. King Henry bowed to me and doffed his cap."*

Bottom right: *The calm before the storm. April 1982 and HMS* Hermes *is going south on her way to join the Falklands Task Force. Ship's company and 'WAFUs' alike enjoying the quiet while they can. (Rod Taylor)*

Far right: *'Wooden wedge technicians' (aircraft handlers), 'Blacky' Blackman (sitting) and Mick Challis (standing), HMS* Albion *off Aden in 1965. (Chris Pugsley)*

CALM SEAS AND 'ROUGHERS'

JOHN FORD: "WHEN aircraft were 'landing on' a consideration arose if the deck was pitching in a big sea. The touchdown point aimed for was some way down the deck, towards the rear end of the flight deck, which might be rising and falling by 30 feet. The decision to land or not, also had to take into account forecast movement of the ship during the next 30 seconds or so. Little 'F' took responsibility for a clear deck and the aircraft's approach path, while 'Wings' looked ahead at the approaching waves giving him an all clear – or not."

Right: *The sea is not always so obliging. An 800 NAS Squadron Sea Harrier FRS.1 approaches HMS Hermes c.1981. Sea Harriers also served with 801, 809 and 899 NAS, all home-based at Yeovilton. (Graham Collins)*

Opposite page: *Lionel Smith: "During Ark Royal's high speed dash to get home, before a dock strike at Plymouth, there was a powerful storm in the North Atlantic sending a heavy swell into the port for'ard quarter. This tended to roll the ship causing the starboard aft quarter to go under violently, causing splits in the hull and a deck to be pushed up against the one above crushing the boat stowed between. The rearmost Buccaneer in the photo has undercarriage damage." (Lionel Smith)*

ROYAL NAVY

HOME FROM THE FALKLANDS

When HMS *Hermes* arrived home to Portsmouth from the Falklands she was greeted by thousands on the dockside, and in countless small boats. Relatives and friends of the ship's company and the air group were there to welcome home the 1,700-strong crew and her complement of Royal Marines, together with survivors of the sinking of HMS *Sheffield*.

She was saluted by flypasts by RAF Harriers, Army Lynx helicopters and FRADU Hunters. As she arrived in Portsmouth, she received a 17-gun salute. The ship carried a scoreboard marking the 46 enemy aircraft shot down by the Sea Harriers launched from her deck. The prime minister, Margaret Thatcher, had been flown onto the carrier by helicopter as it approached port. Captain Linley Middleton was modest about his ship's role in the South Atlantic. "It was all absolutely routine, daily attacks, nothing untoward," he said.

John Keenan was with 801 (Sea Harrier) NAS in HMS *Invincible* during the Falklands War: "The day we handed over to *Illustrious*, ignoring the instruction from the Admiral, it did a steam by to salute us. It approached port side aft, very close to us – going like absolute stink. We could hear that all the engines were engaged by the howling from the funnel. As they passed they streamed a long banner – it read, 'Invincible can eat 3

(Continued on page 132)

Top right: *The 1980s TV commercials for the cereal Shredded Wheat used to say "Bet you can't eat three…" (Illustration by Mike Cole-Hamilton)*

Bottom right: *A view of the incredible welcome from both the dockside and afloat. (via David Rye)*

Opposite page: *Nearly home, back in Portsmouth on 21 July 1982. (Chris McKee)*

POLICE
731

Shredded Wheat'. They were very close and peeled off to port and shot away. I looked down the flight deck, not a soul had moved, silence, no cheering or waving. We were all struck with emotional dumbness. I will take those few moments to my grave, as I am sure will all who witnessed it. I saw my old Buccaneer and Phantom chums aft of the island standing to attention saluting us. Humbling is a totally inadequate description for the camaraderie shared between ships on that day."

Illustrious's Jeremy Browne: "The 'old man' (Captain Jock Slater) was driving her like a frigate!"

Top left: Invincible *streams the Shredded Wheat banner. (Alan Smith)*

Bottom right: *An exchange between the two Sea Harrier Squadrons.* Illustrious *says "Well done 801, swap draft? Luv 809 xxxx" (Alan Smith)*

Opposite page: *How close?!* Illustrious*'s speed can be judged by its wake. (Jeremy Browne)*

HAWKER'S FINEST – THE HUNTER

Former Hunter pilot Mike Layard: "I was an air warfare instructor, and one of the best jobs I had was as senior pilot for the 'Top Gun' school at 'Lossie' – it was terrific. You spent your life pulling huge amounts of 'g' and dogfighting in a Hunter. My collar size went up about three inches. It was a very rewarding time flying that lovely aeroplane – the Hunter at its extremes."

Right: *The other principal mark of Hunter in the FAA was the T.8, with FRADU's WV722 VL-872 seen landing at Yeovilton on 21 September 1974. Other users included 759 NAS at Brawdy and 764 NAS. (Richard Vandervord)*

Opposite page: *Hunter GA.11 WW654 VL-833, of the Fleet Requirements and Air Direction Unit, crossing the south coast whilst up from Yeovilton on 24 September 1985. The GA.11 was a conversion of retired ex-RAF F.4s, all of which had lots of fatigue life remaining. Other users of the GA.11 included 738 NAS at Brawdy and 764 NAS at Lossiemouth. (Robbie Shaw)*

833
ROYAL NAVY

WESTLAND SEA KING

Photos: Jock Alexander

THE ANTI-SUBMARINE SEA King HAS.1, developed from the Sikorsky S-61, joined the navy in August 1969 with 700S NAS at Culdrose. Later developments were the HAS.2, 5 and 6, and air-sea rescue soon became an added role. Other versions were the HC.4 'Jungly' and the AEW-capable ASaC.7. The type served for an impressive 49 years. Squadrons included 706, 707, 771, 772, 810, 814, 819, 820, 824, 825, 826, 845, 846, 848, 849, 854 and 857.

Andrew 'Tank' Murray: "I got my 8,000 hours total, and then my 8,000 on the Sea King. I was privileged to lead the goodbye flypast and the last formation of three Mk 7s on the flight to HMS Sultan (Gosport) for disposal, landing the last one ever on the parade ground there on 26 September 2018. A very emotional event having flown the Sea King for almost 30 years."

Top left: *HMS* Gannet *ship's crest.*

Top right: *Sea King HAS.6 XV632 PW- 03, also of 819 NAS, keeping the rotors running in a challenging landing spot.*

Opposite page: *Sea King HAS.5 PW-826 of 819 NAS based at HMS Gannet, Prestwick on a wintry outing into the Scottish Highlands. The squadron flew Sea Kings from 1971 to 2001.*

2 6
ROYAL NAVY
826

WESTLAND LYNX

THE LYNX REPLACED the Wasp as the standard anti-submarine helicopter on destroyers and frigates. First deliveries were made in 1976 with trials taking place on HMS *Birmingham* and *Sheffield*. No. 815 NAS was established at Portland as the operational unit; the Lynx playing a major role in the Falklands conflict and Gulf War 1. With the arrival of the Wildcat the Lynx was retired on 23 March 2017, by which time 815 NAS had moved to Yeovilton.

Larry Jeram-Croft: "Whilst in the Falklands I remember seeing a Sea Dart missile coming at me – which was amusing! I reckon we broke the speed record for the helicopter as we were at about 6,000 feet, already doing 150 knots. I shoved the Lynx's stick forward, but I've no idea how fast we went. I told Westland's chief test pilot, just before they went for the world speed record in the Lynx, that when you get beyond 200 knots the windscreen wipers just about lift off the screen. He asked me how the hell I knew that – I said I would tell him when he got back."

Right: Newcastle *ship's flight Lynx HAS.2, from 702 NAS Yeovilton, in April 1980. (Chris Taylor)*

Opposite page: *An 815 NAS Lynx HAS.3 ZD257 334, of* Southampton *ship's flight, over Montserrat in July 1995. (Brian Marsh)*

ROYAL NAVY
334
HMS
SOUTHAMPTON

'SHAR' – THE GREAT SEA HARRIER

IF THE SWORDFISH is the FAA's most renowned aircraft type then the Sea Harrier, following its great success in the 1982 Falklands War, must run it a very close second. It saw service with 800, 801, 809 and 899 NAS, all home-based at Yeovilton the FAA's traditional 'fighter town'. Brought into service in 1980, it was prematurely withdrawn in 2006 – a totally unexpected and sudden decision by the government for budgetary reasons.

Right: *Sea Harrier FRS.1 ZA177 with Lieutenant David Smith who shot down a Mirage in the Falklands. He is indicating Dave Morgan's two Falklands kill markings (both A-4 Skyhawks). HMS* Hermes, *Spithead 21 July 1982. (Stephen Wolf)*

Opposite page: *A Sea Harrier FA.2, from 800 NAS HMS* Ark Royal, *refuelling from a VC10 while en route from Nellis, Nevada to Bangor, Maine in 2003. (Robin Trewinnard-Boyle)*

A HARRIER'S FINAL FLIGHT

Photos: Ian Black

All the Fleet Air Arm's Harrier two-seaters were originally built as T.2s/T.4s for the RAF, most later being upgraded by the navy to T.8 standard. They served exclusively with 899 NAS at Yeovilton.

Right and Opposite page: *On 15 March 2006 Harrier T.8 ZD993 723, on the charge of the Defence Aviation and Repair Agency St Athan, made its final service flight. The photographer, Ian Black, was in a Hawk flown by Lieutenant Commander Andy 'Mum' Davies. It was the T.8's penultimate flight, and Ian was there to record the event. After landing and refuelling the Harrier was flown to Culdrose for ground-handling instruction.*

723
ROYAL NAVY

HELICOPTER PEOPLE

Below left: *With the 'chockheads' (aircraft handlers) standing by, a 'Jungly' Sea King HC.4 of 846 NAS lands on HMS* Fearless *in Norway, March 1982, with* Invincible *in the distance. Within a month both ships, the Sea King squadron and the 'chockheads', would all be off to the Falklands. (Rod Taylor)*

Right: *Lieutenant Charlotte Peattie-Primary, 814 NAS Merlin pilot in 2011. "I joined 814 NAS and went out to the Middle East for a period of land-based operations. The squadron then embarked in* Illustrious*. She was part of a multi-national task force conducting an exercise off Albania, during which we operated with various ground forces including Royal Marines. I delivered a large artillery piece onto a hill for them, and was also involved in collecting the Albanian president when he visited the ship. I was part of the crew of the spare aircraft for him and his armed guard pick-up."*

Lieutenant Commander Charlie Fuller, CO HMS Gannet, Prestwick, Search and Rescue Flight 2014–16, in his Sea King. (Tim Woodman)

FAR-FLUNG PLACES – BORNEO

Photos: David Baston

THE INDONESIAN CONFRONTATION of 1963 to 1966 saw the extended deployment of 'Jungly' helicopter squadrons into Borneo. They were to support the army and Royal Marines efforts to contain incursions into the country. Usually, Sibu was the main operating base, with a detachment at a forward operating base (FOB) at Nanga Gaat some 120 miles up-country near the Indonesian border.

Below left: *The second support base in the area was on the island of Labuan. There was an FOB at Bario, high up in the mountains where the air temperature regularly reached 120 degrees Fahrenheit. Hot work on a Wessex Gnome engine for the ground crew at Bario.*

Below right: *Facilities at the FOBs were pretty basic. This is Bario, with the 'control tower' on the left.*

Opposite page: *Wessex HC.5s of 848 NAS deployed ashore from HMS* Albion, *flying 3rd Division troops from Nanga Gaat in 1965.*

ROYAL NAVY
P

AIRBORNE EARLY WARNING GOES ROTARY-WING

When the old HMS *Ark Royal* was retired in 1978, taking the Gannets with it, the FAA was left with no AEW capability. This gap in the fleet's defences was brought into sharp focus in the Falklands. An urgent programme was initiated to fit an AEW radar to the Sea King. This was the only remaining carrier-capable aircraft which could accommodate the Searchwater radar system, housed in a rotating pod. In 1985 849 NAS was re-commissioned to operate the converted Sea Kings. Originally known as the AEW.2, but after a later upgrade this was changed to ASaC.7 (airborne surveillance and control). For a while 849 A and B Flights were renumbered 854 and 857 NAS.

Right: *A Sea King AScA.7 'Bagger' on the later CVS carrier HMS* Ark Royal*'s 1 Spot in 2010. (Robin Trewinnard-Boyle)*

Opposite page: *from 2020 the Sea Kings were replaced by similarly modified Merlin HM.2s fitted with the upgraded Crowsnest radar. This capability is embedded within regular Merlin anti-submarine squadrons. No. 849 NAS was decommissioned on 21 April 2020. (Crown Copyright)*

ROYAL NAVY

'HANDS TO FLYING STATIONS'

BOTH OF THE new carriers, HMS *Queen Elizabeth* and *Prince of Wales* support an embarked air group including F-35Bs, Merlins and Wildcats. Initially F-35Bs have served with 17, 207 and 617 Squadrons RAF and 809 NAS was reformed at Marham in 2023.

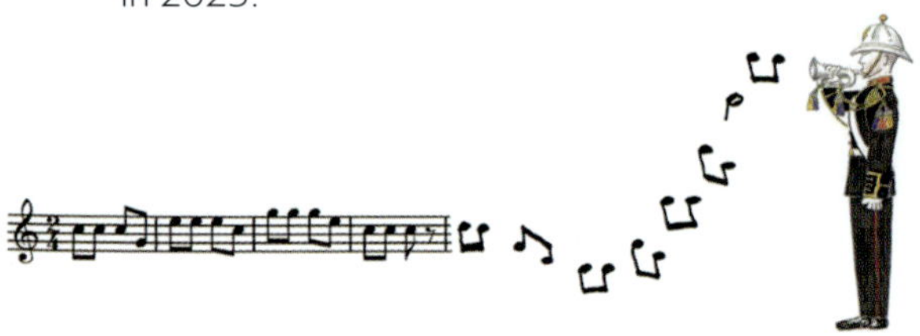

Right: *FAA and US Marine Corps F-35Bs on* Queen Elizabeth *in 2021 during the Carrier Strike Group 21 deployment. (Crown Copyright)*

Opposite page: *An 825 NAS Wildcat HMA.2 firing missile decoy flares at Yeovilton's Air Day, 13 July 2019. (Keith Abnett)*

IN MEMORIAM

Photos: Julie Halford, FAA Memorial Church

THE FLEET AIR Arm Memorial Church, St Bartholomew's at Yeovilton, commemorates all those men and women who have given their lives in both the RNAS and the FAA.

Amongst the many memorials inside the church, one is dedicated to the four naval airmen awarded the Victoria Cross.

From left to right: *Squadron Commander Richard Bell Davies VC DSO, Lieutenant Commander Eugene Kingsmill Esmonde VC DSO, Lieutenant Robert Hampton Gray VC DSC and Flight Sub-Lieutenant Reginald Alexander John Warneford VC.*

ROYAL NAVAL
CEMETERY

INDEX

PLACES

PEOPLE

AIRCRAFT

SHIPS

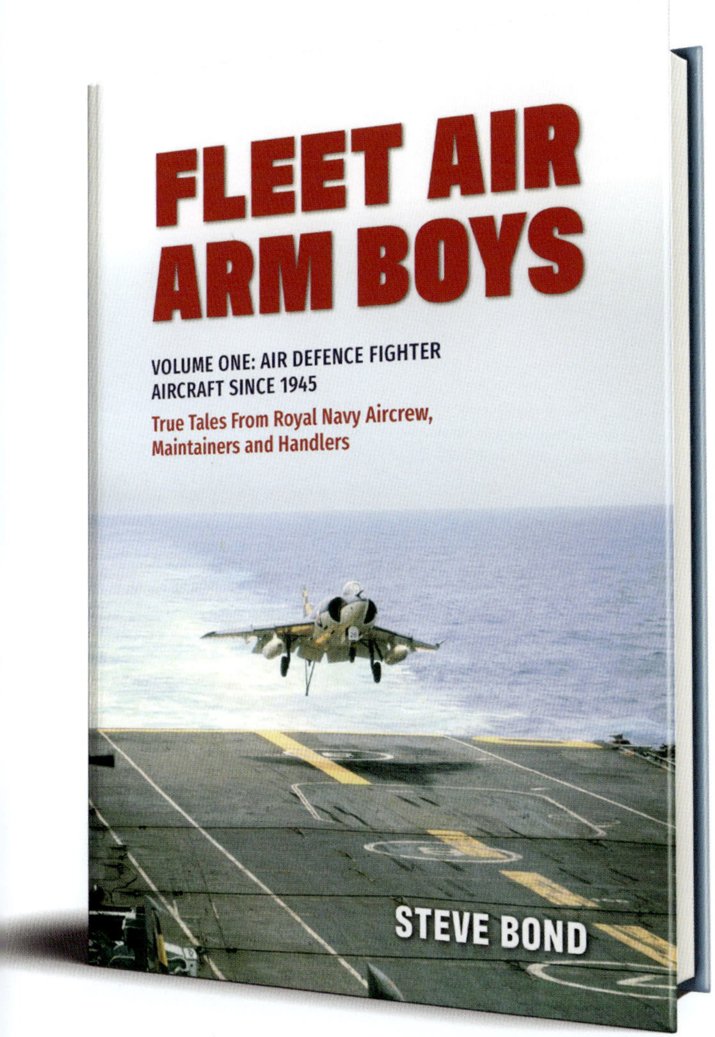

EXTRACT FROM *FLEET AIR ARM BOYS, VOLUME ONE*

AIR DEFENCE FIGHTER AIRCRAFT SINCE 1945

9781911667315
£20.00 | Paperback

By Admiral Sir Michael Layard KCB CBE

The lives of a naval fighter pilot and his observer are hugely challenging, and dangerous at the best of times. As a crew operating from an aircraft carrier brings new dimensions – the ship is always on the move, never where you last saw it, and in rough seas forever pitching, rolling and cork screwing – an even bigger challenge at night of course. All this breeds some very special people with a vital suit of talents.

This book illustrates the whole Fleet Air Arm family, aircrew, maintainers and aircraft handlers alike in their own words. It is an important historical record and tells a story that is long overdue in the telling.

"A meticulously researched tribute to those in the Royal Navy ... A kaleidoscopic collection of photographs, many from personal archives, help to make this book a 'thriller' in its own right."

Pennant

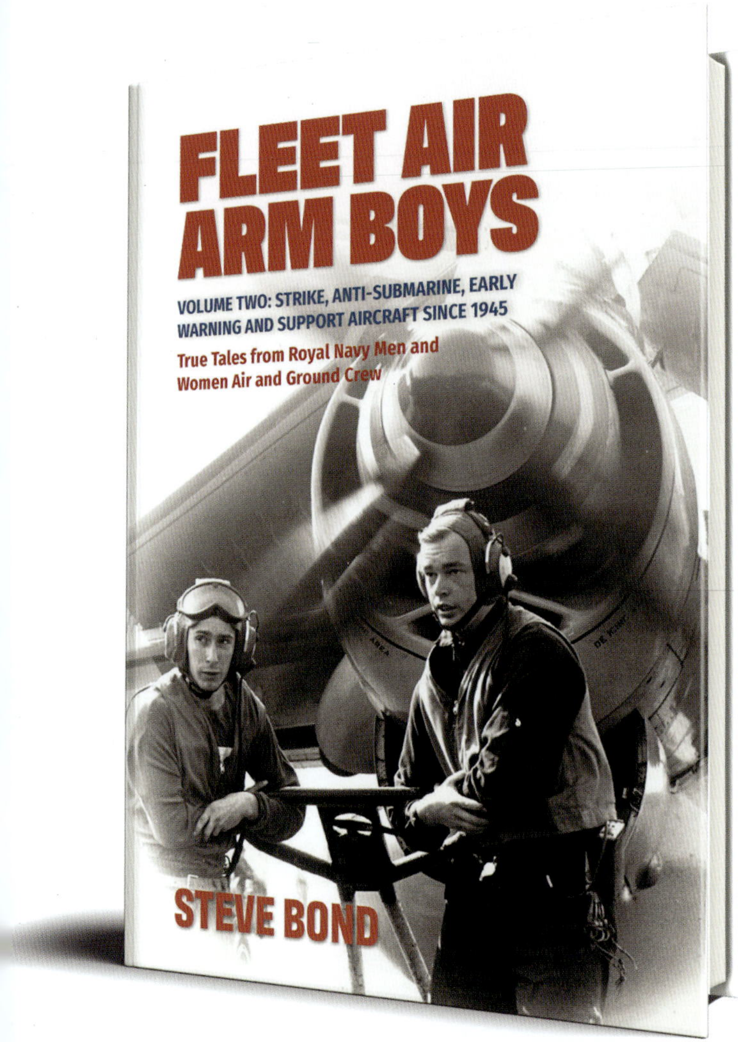

EXTRACT FROM *FLEET AIR ARM BOYS, VOLUME TWO*

STRIKE, ANTI-SUBMARINE, EARLY WARNING AND SUPPORT AIRCRAFT SINCE 1945

9781911667179
£25.00 | Hardback

By Chris Pugsley

Eagle was a good ship. We spent most of our time in the upper hangar, manoeuvring 'cabs' onto and off the flight deck lifts. Jets were big, noisy things with all sorts of sharp points poking at you, and they leaked a lot. As the hangar was designed for Sea Hawks, Sea Venoms and the like, manoeuvring a 17-ton Vixen in it was quite tricky. It was a major challenge operating larger, higher performance aircraft on ships such as *Centaur*. The jets got bigger but the ships didn't, and over the years the deck plates distorted so that when you moved a Vixen it looked as if it was waddling along.

"At a time when fixed-wing naval aviation is coming to the forefront of the RN once again, and catching the interest of the public, Steve Bond's books do a superb job of highlighting the true nature, and strengths, of the Fleet Air Arm."

Fleet Air Arm Officers Association

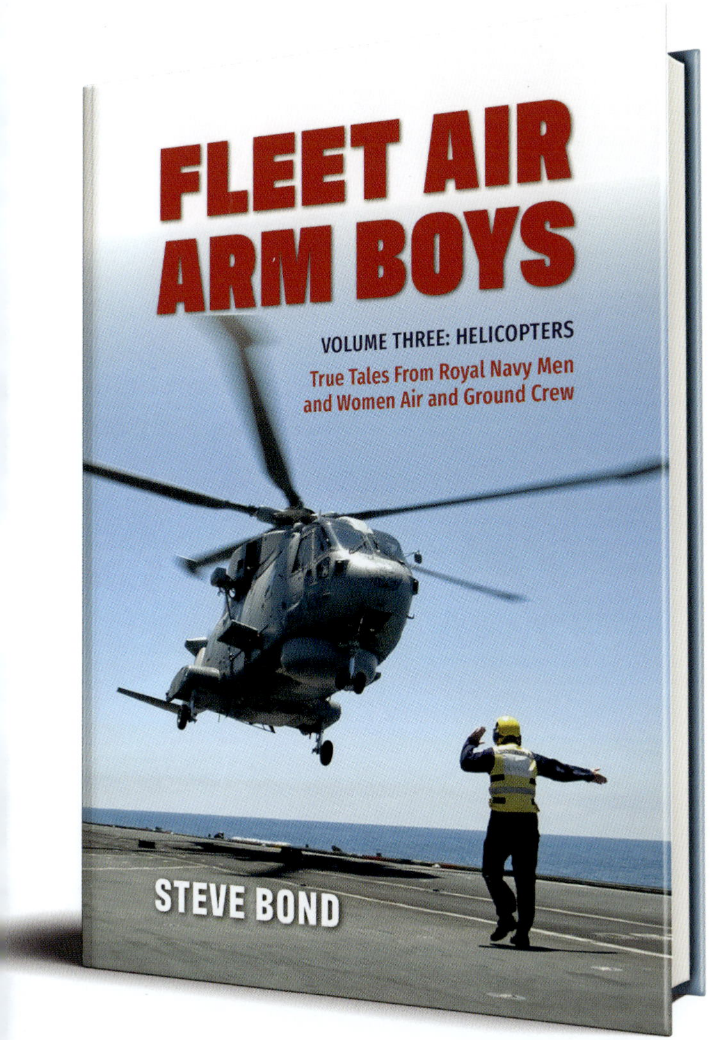

EXTRACT FROM *FLEET AIR ARM BOYS, VOLUME THREE*

HELICOPTERS

9781911667278
£25.00 | Hardback

By Harry Benson

I was a latecomer to the Falklands War, having spent a frustrating month onboard RFA *Engadine*, plodding 8,000 miles south at a sedate 12 knots. Our newly formed 847 Squadron comprised the remaining Wessex crews in Britain, mopped up from SAR units, test pilots, and the Northern Ireland detachment of 845 Squadron (my own first frontline posting). Our squadron split in two; while half sped ahead on *Atlantic Causeway*, our contingent on *Engadine* glided into San Carlos water a week later on 9 June.

The war was about to reach its climax with the final assault on the Argentinian positions in the hills around Port Stanley. I was a newly qualified acting sub lieutenant 'Jungly' three months out of training, and desperate to get into the action. Even if young, foolish and invincible, my first three days on the islands proved that training on Dartmoor and a few weeks in Northern Ireland had put me up to the task. My first mission was a regimental lift of Gurkhas from Goose Green up to the front line beyond Fitzroy. It felt amazing to have been trusted enough to be given responsibility for the lives of my crewman and the dozen troops in the cabin. However the sheer fun of manoeuvring a seven-ton helicopter, just 15 feet above the rolling terrain at 100 knots, was stopped in its tracks by the painful reality check of the still-burning hulks of *Sir Galahad* and *Sir Tristram*, bombed two days earlier.

"This well produced title provides insights mainly from aircrew, but with welcome inputs from the indispensable maintainers... It is a worthy addition to the popular series."

Flypast

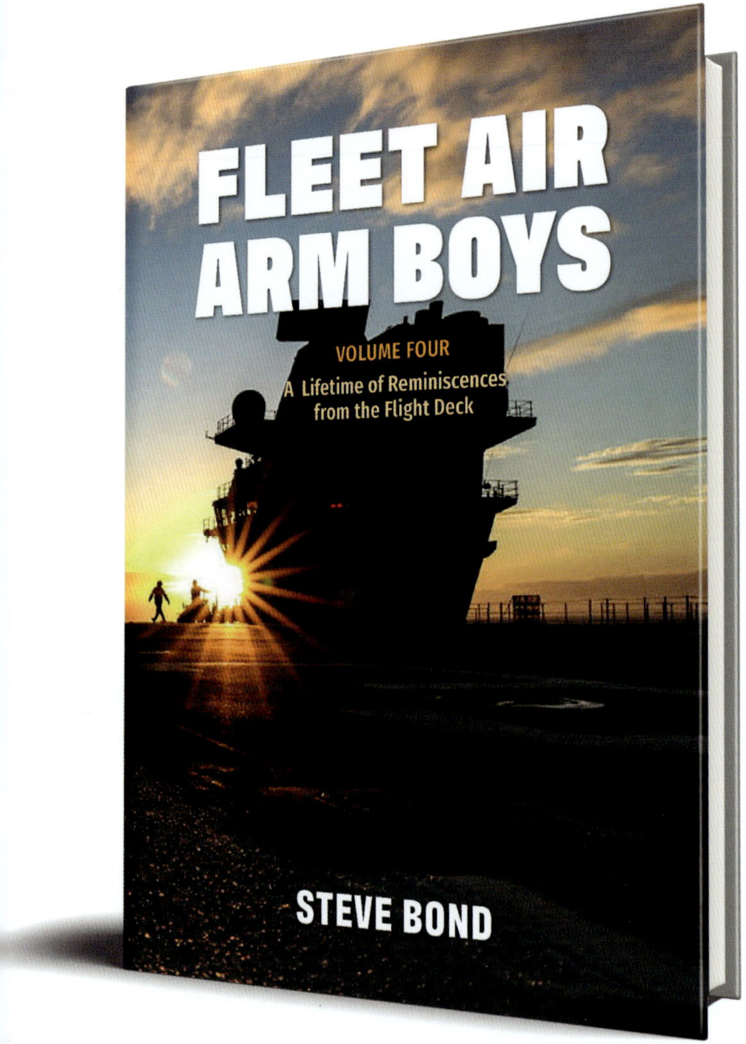

EXTRACT FROM *FLEET AIR ARM BOYS, VOLUME FOUR*

A LIFETIME OF REMINISCENCES FROM THE FLIGHT DECK

9781911667438
£30.00 | Hardback

By Keith Abnett

During the early 1960s I was given the best technical five-year aircraft apprenticeship in the world, and eventually became a fully qualified aircraft artificer. I worked for over 24 years in the most cost-effective air force, which was second to none worldwide. That grounding set me up for life.

I salute all the pilots and observers, both fixed- and rotary-wing, who I came into contact with, as they have my undying respect. They often flew bravely in atrocious and dangerous conditions, on and off a small tin can tossing and turning on raging seas. Their achievements in both wartime and peace should never be underestimated or forgotten.

The camaraderie and friendship I encountered from those days were outstanding and continue to this day. Even now when we meet up and talk the passing years just fall away and memories come flooding back of the squadrons, ships and aircraft carriers we served on. The hustle and bustle of life in the FAA along with the sweat, blood, tears and laughter of service life stay with you forever. I was proud to have belonged to the 'Airey Fairies' – the Fleet Air Arm.

"A particularly thought-provoking read. This volume is more reflective in character and reminded me of a few more sober moments experienced in my eight years in the FAA."

Tony Wilson, contributor to the *Fleet Air Arm Boys* series